W9-CZK-558

EVOLUTION AND THE BIBLE

EVOLUTION AND THE BIBLE

Answers to Crucial Questions About the Origin of Life

By

CORA A. RENO

MOODY PRESS

CHICAGO

© 1972 by
The Moody Bible Institute
of Chicago

All rights reserved.

The use of selected references from various versions of the Bible in this publication does not necessarily imply publisher endorsement of the versions in their entirety.

ISBN: 0-8024-0131-7

Third Printing, 1974

Printed in the United States of America

Contents

Foreword

THIS BOOK is written by a friend of mine for whom I have the highest respect. Cora and I went to Wheaton College together, quite some number of years ago now, where we enjoyed working together in the zoology lab.

Afterwards, when I had been director of Moody Press for several years, Cora sent me her manuscript "Evolution: Fact or Theory." I liked it so well and found it so useful that we immediately went about publishing it. Her later book, *Evolution on Trial*, is equally valuable and a real classic in upholding a biblically oriented view of creation. Now comes this latest and probably most helpful volume of them all.

I heartily recommend this book to students young and old for a better understanding of science and the Bible.

KENNETH N. TAYLOR
President
Tyndale House Publishers

Preface

For nearly twenty years Miss Cora A. Reno's first book, *Evolution: Fact or Theory,* was widely used and accepted as a valuable laymen's guide to a biblical position amid evolutionary theories. During that time the book sold over 200,000 copies in its twenty-one printings.

Gradually, the information contained in *Evolution: Fact or Theory* became out-dated and in need of revision. Partly in response to that need, in 1970 Miss Reno wrote *Evolution on Trial* for the student and interested layman looking for a thorough, scholarly response to the continuing onslaught of evolutionary hypotheses.

Now in *Evolution and the Bible* comes a popularly written abridgement of *Evolution on Trial* meant to fill the need for a briefer, general study since *Evolution: Fact or Theory* went out of print. For a more detailed study of the information in *Evolution and the Bible,* the reader is directed to Miss Reno's *Evolution on Trial* (Chicago: Moody Press, 1970). For other texts on evolution, see the bibliography on pages 185-89 of *Evolution on Trial.*

The Publisher

1

Is There Conflict Between Science and the Bible?

For many people this is not a settled question, yet it is one that needs serious consideration. Every day scientific inventions and discoveries are continually changing the world. As science changes, should the interpretation of the Bible also change? Are the two in harmony or are contradictions present? Any belief in God has to include His being a God of truth, but contradictions can never originate from a single source of truth. After this step of reasoning must come a judgment about the validity of the Bible.

Studying the Bible eventually leads to the conviction that it is the very Word of God. When we accept the Bible as originating from God, our next step is to believe what it says about the source of all science.

God is that source. According to the Bible, all things were made by Him. Since He made all things and set into operation the laws by which they function, things must all be in harmony. No matter what branch of science one is studying, when rightly interpreted, it points back to God and the truth which originated in Him.

We believe the only sound premise to work within holds that both the Bible and all science come from God. Since this common source is ultimate truth, no contradiction can exist between them.

In this book we will try to be fair and honest in examining the viewpoints of those who do not agree with our premise. Since this book deals almost entirely with evolution, concentration will be on the fields of science which are in some way connected with this subject. You may be surprised to find that this includes almost all of the branches of science!

2

Is Evolution Still a Problem?

INDEED IT IS to a great many people. Evolution is a subject that is still debated, studied, and argued today, as much as or more than it was a hundred years ago. The origins of the universe, of the earth, of life, and of the variations among living things are the topics involved in this continuing debate.

Association with young people and interest in their problems leads to a concern with this issue. A young person finds that as soon as he opens his textbooks in school, at either the elementary or high school level, he is faced with the theory of evolution. This theory is usually presented as a fact, thus appearing to be the only explanation to account for origins and variations.

The purpose of this book is to show that the theory of evolution is not an established fact. The word *theory* implies more than one side to examine; and this is what we will be doing. Organic evolution is usually the only concept of origins presented in elementary and high school science books and, since pupils at this learning level are not able

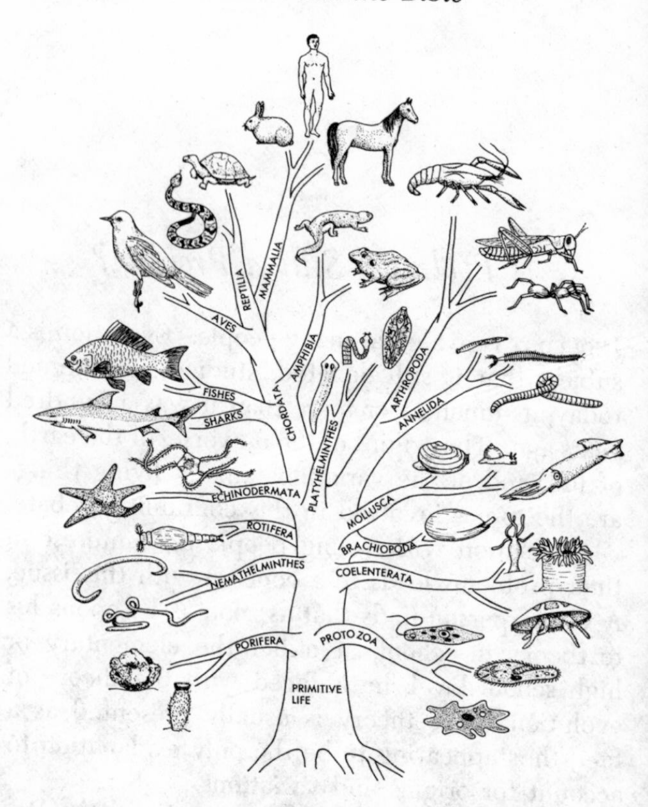

Figure 1

Phyla of the animal kingdom arranged like the limbs of a tree. This picture is typical of the concept most high school books present concerning the theory of evolution. It is found in *Biology For You* by B. B. Vance and D. F. Miller, 1963 (used by permission of J. B. Lippincott & Co.). They say that the animal phyla can be arranged like the limbs of a tree. The limbs closest together are thought to be most closely related.

to evaluate the theory, it is generally accepted without question. In advanced texts, the limitations of the various statements supporting evolution are much more freely admitted. Upon examination, it will also be seen that the points given to support evolutionary theory in college courses differ somewhat from those given to younger students. This book will be dealing only with evolution as it is taught in high school.

Talking about the word *evolution* requires some agreement about the meaning of the word as used in this book. The word itself simply means "change" but more is implied by the way that we and the texts to which we refer will be using it. An organic evolutionist usually says that by change *all* living forms in the world today have been produced from one or a few common ancestors. The process of evolution thus refers to the succession of changes by which organisms pass from the simple to the complex. Often, in the books we will be considering, the term *amoeba-to-man evolution* is used. This book will do the same.

When we say that it is generally believed that *all* living things in the world are related, we mean just that. Evolutionists maintain there was change from one group to another so that each plant and animal in existence today is related to each other one. One much-used high school text says it this way: Evolution was once hotly debated, but is now a well-established fact. Because of evolution, all present-day plants and animals have a similar history, their beginning going back to the origin of life itself. By slight changes, over thousands of years, the descendants have changed to become the plants and animals of today.

We cannot agree with this position nor with a book that says that all reputable scientists agree that the evolution of life on this earth is a well-established fact. Remember that evolution is *not* a fact, but rather a theory. Theories usually have some value *if* they encourage you to study both sides of the question. Let this book be a challenge to do this. Do not accept any theory because you read it in a book, hear it taught by persuasive people, or see it presented in an ingenious way. Do some critical thinking; and above all, compare everything with what the Scriptures present.

3

Does Similarity Show Relationship?

THE TEACHING that things are related if they are similar is one of the chief arguments used to support the theory of evolution. Since this concept is so basic to the theory, we will consider it first and then refer to it later a number of times in various forms.

Many high school texts use pictures to teach the idea that similarity shows a common ancestry in the remote past. Some give a sketch showing how the arm of a man, the flipper of a whale, the foreleg of a horse, the foreleg of a frog, and the wing of a bat and a bird are very much alike in part of their bony structure. It is then stated that these appendages of the different animals are alike because they have a common origin. This concept is then pushed to the point of saying that all living things are related. The evolutionist assumes that all living things come from one simple form of life, and therefore it is to be expected that likenesses will exist.

How can a creationist account for the likenesses

found in living things? Of course we admit that they do exist, and that some of them show different degrees of relationship. If we take the case of man, we notice that the Caucasians look more like each other than like Negroes. Within the Caucasian race, Scandinavians look more like each other than they look like Italians. Within the Italian nationality, members of the same family often show a family resemblance. Identical twins within a family will look even more alike than other brothers and sisters because, having come from a single egg, there is a closer relationship.

In this example we see that the more closely related men are by race, nationality, family, or twinning, the greater is the similarity. Believing that all men came from the originally created pair, Adam and Eve, we have to agree that their offspring show similarities and yet very obvious differences, too. The same observable differences are true within other groups and, therefore, this is not just an evolutionary argument. From an originally created pair of ancestral horses, or dogs or chickens, could come a variety of horses, or dogs or chickens. However, we also note that as we follow up this idea of relationship shown by similarity, we eventually come to great gaps. There is not just one "missing link" in the theory of evolution but literally thousands of them. They exist among all the major groups of plants and animals.

Is not the first thing that comes to your mind when thinking about the differences and yet the likenesses that we see, that they could be the result of a common plan in the mind of God the Creator? In fact, it is logical that God would use the same general plan for many different plants and animals. When living things have the same physical functions and must live under similar conditions, it is natural that the equipment for these functions would be similar. God logically would give the same structure to animals who were to walk the same earth, breathe the same air, and eat the same food. The great Designer might have used another general plan for creatures that were to fly in the air and another for those that were to live in the water.

When we say that God has all power, we realize that He could have made every single plant and animal different from every other one.

This would have resulted in many weird combinations of numbers of legs, ears, eyes, and heads. It is not a question of what God was able to do, but what He actually did. Creationists believe that God created large groups of plants and animals according to a general plan and gave to them certain similarities at the time they were created. Individuals with great differences developed within these groups, as we have seen is true of man.

Because similarity is such a basic argument to the theory of evolution, let us look at some examples of its weaknesses.

First: Certain parts of a plant or animal may show similarities and other parts show just the opposite. The similarities in the skeleton of the Tasmanian wolf and the dog are used by the evolutionist as evidence of close relationship. However, the Tasmanian wolf is a marsupial, carrying its young in a pouch, whereas the dog is different. Apparently, the two are not closely related. The bill and webbed feet of an Australian platypus make it resemble some birds, and yet it is classed as a mammal because it has the very unbird-like characteristic of feeding its young from milk glands.

Second: Parallel mutations cause another difficulty in accepting similarity as evidence of evolution. In a zoo at the same time might be born an albino deer and albino bear. Because they both have a superficial resemblance, no one would suggest that this deer and bear are any more closely related than other bears and deer. The change in germ cells which caused these animals to be albinos just happened to appear in the two animals at the same time.

Third: Likeness in appearance is not a reliable guide to kinship. The male and female black widow spider look so different that an untrained person would not even recognize them as belonging to the same species. The same is true of the red-winged blackbird.

Fourth: Creatures that look very much alike sometimes are really so different that they will not interbreed. By looking at them, an observer can hardly tell the Arcadian from the Least flycatcher, and yet this likeness is only superficial.

Fifth: Some immature forms look so different

from the adults that they cannot be recognized as belonging to the same species.

In summary, similarities can point to creation by God, the great Designer, or to the fact that the plants and animals really have had a common ancestor. Although sometimes it is true that unrelated animals may appear similar, this is not good evidence for evolution. It is not a consistent argument, because certain parts of some animals are similar whereas other parts of the same animals are very unlike. Internal parts may show one thing and external parts another. Some similarities are the result of parallel mutations and not kinship. Similarity is often superficial: some animals look alike but will not interbreed. Animals of the same species which would be expected to look similar may differ greatly because of age or sex. Rather than evolution, creation is a good explanation for similarity.

4

Do Fossils and Geology Indicate Evolution?

IF EVIDENCE for the theory of evolution is to be found anywhere, it should be in the rocks and fossils. This is objective material and demands explanation. Since rocks and fossils are such crucial subjects, study them with extra care.

There are a number of different kinds of plant and animal fossils. They may be the whole organism or just a part seen as an imprint—petrified or preserved in rock or other substance like ice, tar, or amber. The most common kind was formed under water, as sediment settled to the bottom and covered the plant or dead animal.

The study of fossils and geology is a difficult one, and you may find things that will be new to you. Do not be afraid to study them; and then try to evaluate the material to see if you can learn more of the way in which God has worked in our world in the past. In some ways, this study is like space exploration. In both fields, the facts and figures of last year or even last month often need revision. As we continue to develop new dating methods and learn more about the rocks and fossils, we must update our theories about plants and animals that lived in the past and about their history down through the ages. This is the way a scientist must work. He sets up theories and tests them to see if they meet existing facts. If they don't he often has to make some changes in his theories. This should be to his credit rather than used to bring him into disrepute. Cautious analytical thinking is especially needful in the areas covered in this chapter.

As we begin to think in detail about geology, we need to be familiar with the term *stratigraphic column.* You may see one in a simplified form as you stand and look at a hillside where erosion has cut away the soil exposing layers of rock. Of course there are many exceptions, but usually the oldest ones are deepest, and the ones formed very recently are nearer the surface. At no one place in the earth is there a complete stratigraphic column; but parts from one place can be filled in (by a person who really knows his geology) to fill the gaps at another place, so that we have a fair idea of what a complete one would be like.

There is much work involved in setting up a stratigraphic column and making a geological timetable such as will be found in most books. Scientists know the average rate of time for laying down different types of sediment that make up the rocks. This knowledge and other tests are used to calculate the age of a particular rock. It is realized that from time to time the factors that control these rates of rock formation may undergo changes. However, no evidence indicates that the rate was ever very different from the present one. Thus, the approximate dates given in the average high school text for the various eras and other time periods represent reasonable deductions based on average rates of deposition. Geological time has been divided into various subdivisions in much the same way as historical time has been divided into segments based on what man did at that particular time. We speak of such historical periods as the Dark Ages and the Renaissance. These periods are part of the Christian era. In much the same way, geologists divide the earth's history into great eras. Some of these eras are divided into periods and then subdivided into epochs.

Although we realize that there is no way to get an exact time to assign to the various geological time periods, most geologists feel that they have close enough accuracy to be generally helpful. Of course we know that in connection with geology and fossils there are many problems for which we do not have adequate answers. However, the majority of scientists feel that we should not hesitate to use terms running into millions and billions of years. Even before the use of radioactive dating methods, people had seen many things that made them think the world was very, very old. We cannot consider all of them, but let us list a few.

Australia's Great Barrier Reef and coral islands were gradually built up from secretions of tiny animals, many about the size of a pinhead. As colonies of these animals died, innumerable trillions of skeletons gradually formed the reefs and islands.

In Yellowstone National Park there is a two-thousand-foot cliff. A cross section of it shows eighteen petrified forests arranged in linear fashion, one on top of the other. It would have taken an immense amount of time for each forest to grow to maturity, be destroyed by volcanic ash, and then have a succeeding forest grow on top of it.

Calculating the rate of the formation of sedimentary rock also makes us think that the earth is very old. It is assumed that the rate at which it was formed in the past is at least somewhere near the rate at which this rock is being formed today.

Measuring the rate at which erosion wears away land areas gives us another measure of the time the earth has been in existence. Of course this does not indicate the time of creation, but we can tell a little about changes by seeing the vast amount of time it has taken for the rocks under Niagara Falls to wear away—in fact, for the falls to be found in their present location. The Appalachian Mountains are very slowly being worn down, and it must have taken an almost unbelievable amount of time for the Colorado River to cut out the Grand Canyon.

As a person stands in the Mammoth Cave or Carlsbad Caverns and looks at the huge stalactites and stalagmites, he must realize that it took a long period of time for their formation, which is due to water containing limestone slowly dripping from the roof of the cave.

Many high school texts today still list the increasing concentration of salt in the ocean as one of the reasons for believing that the earth is millions of years old. There is reason to think that at the time of creation the oceans may have been composed of fresh water. As rain fell and ran into the ocean by way of streams and rivers, it took with it salt and the other minerals leached from the soil over which it flowed. As the water evaporated, it left behind the salt, thus increasing the total amount that the ocean contained. This increase in concentration is still taking place. Because conditions are always changing, the rate of increase likely would have varied from age to age. Therefore, it cannot be used to arrive at any absolute date, but it does help us to realize that the earth has a great age.

Gypsum layers hundreds of feet thick are found in the Southwest. At the present rate of deposition, it would have taken tens of thousands of years for these to form.

These things just mentioned—coral islands, forests on top of one another, sedimentary rock formation, erosions, stalactites and stalagmites, salt concentration in the sea, gypsum layers, and many other factors too—lead us to infer that the earth is millions of years old. Because such things as rain, wind, and temperature may vary greatly from time to time, and natural processes may be slowed or speeded up, any dates arrived at by these methods are very crude. In recent years, more accurate methods of dating have been discovered and are now being used. Carbon 14 is one of these methods. Cosmic rays in the upper atmosphere act on nitrogen to produce radioactive carbon, which is different from the more common carbon 12. While plants are manufacturing food, this radioactive carbon is taken into the plant, and at its death a certain fixed amount is present. This is also true of animals which feed either directly or indirectly on plants. After death, of course, no more enters the plant or animal, and what is there starts to disintegrate at constant rates. We can tell how long ago death occurred by measuring the amount left in a plant or animal.

When first discovered, the carbon 14 method of dating was used on materials whose ages were known. This was done to establish the validity of the tests and usually proved to be well within the limits of error allowed for scientific experiments. For example, a funeral boat taken from a tomb in Egypt was tested. The known date of the boat and that given by the carbon 14 method showed good agreement.

You can see how this method of testing dates has been of incalculable help to the scientist. Suppose he were studying the Cliff Dwellers of Arizona and after much study came to the conclusion that they had lived 12,000 years ago. Now suppose he took some tool or weapon found in the cave and tested it by the carbon 14 method. If the result of that test showed about the same age, it is obvious this would be very helpful in confirming other estimates.

Another recent discovery has led to the method called "fission tracking." This is especially useful if one has volcanic glass containing uranium 238 even in traces as small as one part in a million. The uranium, being radioactive, divides at a regular rate, leaving "fission tracks" in the glass. Under a microscope, these resemble tiny grooves which can be counted and thus can provide a measure of the time that has passed since the glass cooled. As with the other tests, there are several factors which can cause an error in the date at which one arrives. Fortunately, these factors are different from the sources of error which enter into the use of some of the other tests; therefore, one type of test can be used to check against another.

The potassium-argon dating method is another one that involves a radioactive substance. In this case it is the element potassium. When it decays, it forms calcium and argon. This test can be used on material that is very old, since the potassium decays slowly. When employing this method, a scientist uses an apparatus called a mass spectrometer to measure the amount of argon in the rock. Since he knows the rate at which the argon is formed, he can tell the date at which the first argon was produced, which is likely the date of the rock's formation.

Another common method used to estimate the age of the rocks utilizes uranium 238, which disintegrates very slowly. It forms an isotope of lead which is easily recognized because it is different from the regular lead with which we are familiar. In four and one-half billion years, half of this element transforms itself into lead. If one should find in a pocket of undisturbed rock a certain amount of lead and of uranium 238, it would be possible to estimate how long it took this amount of lead to be produced. By this method, some rocks have been shown to be many millions of years old. This is thought to be a fairly accurate minimum age.

Many agree that in this field of assigning dates to various material, much work is yet to be done. In addition to the radioactive methods mentioned, we could call attention to the use of thorium, strontium, and rubidium. Others will likely be discovered. These methods are more accurate than the ones listed in the first part of the chapter, because they are not influenced by ordinary environmental changes of heat, magnetic and electrical fields, vacuum, and light. As information about the age of the earth is checked by one method against another, some of the findings are always substantiated and others have to be revised. Accepting these old dates should in no way be construed to mean that one is accepting the theory of evolution, "giving ground to the evolutionist," or setting up science as a sacred cow. It is strange that many people will readily accept scientific discoveries in the fields which give them a more comfortable life by improving medicine, food, ways of transportation, yet will reject equally well-authenticated discoveries in the radioactive field.

Concerning evolution, the significance of our discussion about the age of the earth and rocks lies in the fact that they contain fossils. Fossils are our best key as to whether evolution has or has not taken place; and if we know something of the age of rocks, we also will know something of the age of the fossils they contain.

Whether written by a creationist or not, most books on geology will contain information about a geological time chart based on the stratigraphic column. As mentioned before, various periods of time have been given names and assigned approximate dates. It is significant that on the deepest rocks are found no signs of life. In the next few higher levels can be found remains of plants and animals similar both to the ones living today and to ones that are very different. Some of the fossils are of living things that became extinct millions of years ago; and some are of those that became extinct in comparatively recent times. Some are never found in the same strata of rock with man. Others, like the mammoths that lived maybe ten thousand years ago, were hunted by man; and the spear points that killed some of these mammoths have been found in the fossil bones.

In some layers of rock, we find enormous numbers of fossils, even millions and billions in a very small area. We have to recognize that certain forms of life were once more abundant than they are now. Since some abundant forms are found chiefly in certain layers of rock, they are spoken of as the dominant form at the time that rock was formed. Because there are differences in the forms found in one layer of rock and those found in another, we know that changes have taken place throughout the ages.

One author of a high school text says that it is "a very curious circumstance" that the oldest rocks are so barren of fossils. To a creationist this is not curious, because we know that living things were not created at the very beginning. However, when we come up to higher, younger rocks, those of the Cambrian layer, we suddenly find all kinds of fossils. This is sometimes called "the Cambrian explosion." These rocks are generally believed to be about half a billion years old, and in them are found representatives of most of the major groups. Some of the vertebrates are not found here, but there are jellyfish, sponges, worms, snails, and trilobites. Instead of discovering them evolving one from the other, they are found suddenly appearing in the rocks with their complex systems and organs fully formed and functioning. To the creationist this is the picture of creation. To one who leaves biblical, miraculous creation out of his thinking, there is no satisfactory answer for the quantity, variety, and complexity of the Cambrian fossils. The explanation for this is found only in Genesis. It would appear that at the time of creation, God brought into existence organisms fully formed, with all the parts and abilities needed to carry on life in the habitat in which He placed them.

In these Cambrian fossils are not found any links between the large groups, or phyla; but also there are unbridged gaps between classes and orders within these groups. Gaps unfilled by plants or animals are one of the chief weaknesses of the theory of evolution.

We have said that among the fossils in the very old Cambrian rocks there are representatives of all the major groups found living today. This includes even the simple ones that, according to the theory of evolution, might have been expected to die out. Simple and complex organisms that can hardly be distinguished from the same kinds of simple and complex ones found today, lived millions of years ago. According to the evolutionary theory, one would expect to find whole groups of fossils of simple organisms. Instead, they became extinct as they were replaced by newly evolved ones which are supposed to have been better adapted to a particular set of surroundings. Most of the books we are considering contain the idea that organisms became extinct because ones better fitted to live in a particular environment evolved and replaced them.

In rocks younger than the Cambrian types, the same quantities of fossils do not all appear at the same time; rather there is a succession of appearances, usually set apart by a gap from the last and older ones. They do not show an amoeba-to-man picture. This succession fits in with the biblical creation picture, in which various kinds of animals appear on the fifth and sixth days of creation, as one event followed the other. Concerning the animals, Genesis 1 says, "God created great whales, and every living creature that moveth, which the waters brought forth abundantly . . . every winged fowl . . . cattle, and creeping thing, and the beast of the earth after his kind: and it was so" (Genesis 1:21, 24).

If the creation of all animals was not instantaneous, but one event followed the other, it is reasonable to suppose that the creation of animals within a group also would be done in orderly succession. If God created "fowl that may fly" and later "cattle and creeping thing and beast"; then within the former group, He may have first created ostriches, later eagles, and still later owls and sparrows. Each group was told to be fruitful and multiply. Some scientists think that the various kinds of animals did this multiplying to the extent of becoming the dominant form at a certain time. This does not make it necessary to connect a special era or period on the geological time scale with one of the creative days. However, it is known that fish were so predominant during the Devonian period that it is called the Age of Fishes; the Mesozoic era is called the Age of Reptiles; and the Cenozoic is known as the Age of Mammals. After their particular period of greatest fruitfulness, as new groups were created, some of the animals from each of these ages became extinct, although some persisted and are still living today. The latter, of course, is true of man, who was last to be created.

Both the Bible and science agree that man was the final creation. His fossils are found only in the upper crust of the earth, and he is the highest form of life, because he was created in God's image. As God formed each new group, gaps existed between it and ones already created. These gaps show in the fossil record and are recognized by all as one of the chief weaknesses of the theory of evolution. Even Charles Darwin said that the lack of intermediate and finely graduated fossils to form an organic chain from the simple to the complex, was one of the most obvious and serious objections against his theory of evolution.

This great weakness, gaps in the fossil record, is recognized by the writers of most high school texts. This inability to sustain amoeba-to-man evolution is covered by phrases about the large groups of plants and animals, such as, "These suddenly appear in the rocks"; "No forerunner of this organism is known"; "They appear as if from nowhere"; "There was a burst of this form." Thus we find that those people who begin their thinking with a philosophy prejudiced toward evolution are left with no satisfactory answer to this most crucial problem.

If evolution were true, one would expect to find many connecting links between the different groups, showing some of the changes as one kind turned into another. Very few animals are usually listed as evidence that the gap between two distinct groups can be bridged. However, this evidence is scattered and not at all convincing. Most of the books we are using list only the euglena and the archeopteryx.

Because it contains chlorophyll, the euglena is sometimes called a link between plants and animals. We agree that it has characteristics of both. However, the argument is largely an objective one present in the mind of the reader that "similarity proves relationship." As shown earlier, this is not valid.

We also agree that the archeopteryx has some characteristics of birds and some of reptiles. We think that this similarity does not show that it is a connecting link between the two groups. On the evolutionary scale there is an ascending order from simple to complex reptiles and then from simple to complex birds. For this animal to be a connecting link, it should have characteristics of the most complex reptiles and then of the simplest birds. This is not the case. Instead, the archeopteryx has characteristics of some of the most complex birds, for its feathers are of the most highly developed type. Another weakness in this argument is the fact there is no other animal to connect it to a reptile-like ancestry. It suddenly appears as it is with no indication that it evolved from some other animal. Thus it seems expedient to conclude that it is a distinct animal with enough birdlike characteristics to warrant classifying it in this category.

In addition to the two we have mentioned, other plants and animals have been suggested by the evolutionists as spanning gaps between groups of fossils. However, as one studies each individually, he will find each to be a separate and distinct organism. An evolutionist often explains the gaps in the geological fossil record by saying that the fossils either were not formed or as yet have not been found and studied. He thinks that organisms once existed to give a complete picture from the simple to the complex.

It seems much more logical to suppose that the missing links never existed, and that the first of each group was specially created as a miraculous act of God. The same reasoning applies to the unbridged gaps in the formation of the parts of the various animals. Such parts as wings, arms, legs, and eyes are found fully developed, functioning, and useful. That this could have happened by chance does not seem reasonable.

Let's think back over what has been said in this chapter. In addition to the older methods of dating the rock strata and fossils, more accurate radioactive tests are now being used. When at all feasible, more than one method should be used on a particular sample. The preponderance of evidence makes it seem reasonable to accept a date of four or five billion years ago for the creation of the earth. We also conclude that the creation of one-celled plants and animals may have been during the pre-Cambrian period, but they surely were well established in the Cambrian, along with representatives of the major groups. In rock layers that form a vertical column above the Cambrian, we find strata with fossils arranged in the order in which it appears that God created them. Both the Bible and science agree that water life was formed first and that various forms of plant and animal life appeared progressively, with man being created last. Since their creation, living things have descended with considerable modification. Some became extinct after having had an age of dominance. Between the groups, great unbridged gaps are found. This is in harmony with the biblical account of creation.

5

What Should We Think About the Horse?

All the books we are examining give some account of the horse and use it as evidence to support the theory of evolution. Therefore, it is especially important material to examine. The accounts in most books are very much alike. They tell of an animal, called the Eohippus, about the size of a small dog and estimated to have lived about fifty or sixty million years ago. It had four toes on each front foot and three on each hind foot, each toe with a tiny hoof. Such parts as the teeth, muzzle, and brain size are then described and compared with the modern horse. This is followed by such statements as "All these differences are the result of gradual changes," "Each change in the horse must have been very slight," and "Natural selection guided the direction of their evolution which was brought about by gene mutations."

All but two of the high school books we are examining put this section about the horse in a chapter clearly labeled as dealing with evolution. Most of the books arrange their pictures as a series portraying an increase in size—from the small Eohippus to the large horse of today. All the books include some pictures of the toes and legs, and most of them also picture the teeth and skulls.

That there has been much change within the horse "kind," all will agree. However, scientists are not agreed on how much change there has been and how this should be interpreted; but scientists do have to recognize that it is all change *within* a group. It is not the kind of change that would cause one major group to become another major group. Therefore, it is not *evolution* in the sense in which we are using the word.

Many skeletons of horses and horselike animals are found in museums. These must be studied and fitted into some picture. Two general interpretations will be found among creationists. Neither of these should be considered as favoring evolution. There are conservative men who think that the horse does not show a series either in time or space. They think that within the horse series there may be three, four, or even five separately created kinds. From these have come the variety we see in the world around us. The Bible surely does not teach that each different kind of ponies and horses as found today indicates a separate creative act. Many think that those existing today have descended from the few created ones mentioned above.

Even a larger group of conservative scientists thinks that not only horses like the Arabians and Percherons, but also others like the zebra and ass, came from one created kind. Genesis seems to allow room for this when it says that God created great groups as "beast of the earth . . . cattle . . . everything that creepeth . . . fish of the sea, and . . . fowl of the air." There is no indication that each of the hundreds of thousands of distinct species were created separately. Fossils of the horse and even changes that are going on today would indicate that as much change as is seen in the horse is very possible.

Since some nonevolutionists think that several ancestral horses may have been created, while others think that there may have been only one ancestor to the horse and to several closely related animals, can you see the reason why men are constantly studying to try to supplement our knowledge in this field? All questions are not settled in the minds of creationists and there is still room for *your* study in this area.

Our conclusion about the study of the horse must be that it does show change, but these changes do not fall within our definition of evolution. No one should be disturbed by change or diversification that has taken place in horses any more than about the change that has taken place within other groups of plants and animals. Some people are disturbed because the change in horses is an argument usually given by the evolutionist in favor of

ıis theory. This should not be, but the one is usu-
ılly associated with the other. Some people think,
herefore, that because they repudiate evolution,
hey must also repudiate the diversification seen in
he horse. However, it is not the *kind* of change
hat could ever cause one major group to become
ınother major group, which would be necessary if
volution were true.

6

Should I Consider Classification Good Evidence for Evolution?

CLASSIFICATION, OR TAXONOMY, is a very useful branch of science, because it is easier to work with things that are arranged in an orderly manner. In connection with living things, classification is a variation of the idea that similarity shows relationship. It is stated that the more nearly alike plants or animals are, the closer they are related in their evolutionary ancestry. The evolutionist says that classification leads him to the arrangement of large groups of plants and animals in the order of their probable appearance on earth. He goes on to say that members of the same species are closely related, having evolved from a common ancestor and that similar *species* are grouped together into a *genus.* Similar *genera,* with a more remote common ancestor, are placed in a *family;* similar families make up an *order;* similar orders make up a *class,* and similar classes a *phylum.* The highest category is a *kingdom,* which is composed of similar phyla.

If one starts with the proposition that evolution is true, this brings him to the conclusion that jellyfish and man have very little in common because their common ancestors existed in very remote geological times. The common ancestor of a frog and man are not so far back in history, so they are nearer alike; and the ancestors of a man and an ape would be even more recent. It is not said that man came from the ape, but rather that somewhere in the past they had a common ancestor. This ancestor is said to be more recent than that shared with the jellyfish or frog and, therefore, they have more likenesses.

We all have to agree that things can be classified, but is this in any way evidence for the theory of evolution? After you think about this question a little, you will come to the conclusion that classification is not evidence for, or against, either evolution or creation! In either case plants and animals are taken as they are found; and, working with what is given, one proceeds to classify them. Every day this arrangement is done with our money, stamps, buildings, furniture, and even the words in a dictionary. For example, money can be divided into paper and metal. Metal coins can then be subdivided into ones that are more alike on the basis of size, value, or the metal from which they are made. The fact that ones which show special likenesses are put into certain groups does not mean that they come one from the other. Neither does it show that the same person designed them, although this might well be true.

The branch of science which has to do with classification is an important way of systematizing biological knowledge. However, we must realize that it has little value in pointing either toward evolution or creation. It is a manmade scheme, and there is no classification in nature. If it pointed either way, it would be toward creation and design. Plants and animals are similar because God created them that way. It was logical for Him to use the same general plan for a number of different organisms, rather than each individual being created different from all others. Using classification as proof for evolution is not true scientific evidence.

7

What Should I Conclude About Biochemistry?

CERTAIN CHEMICAL RESEMBLANCES among living things are also pointed to as further evidence for evolution. Similarity in the composition and action of various substances and secretions within living things shows genetic relationship, the evolutionist states. We will discuss in this chapter a number of substances that advanced texts would not put under this heading, but we do so because of the high school texts we are explaining.

Enzymes and hormones are included in one line of investigation. Pepsin and trypsin are found throughout most of the animal kingdom. This fact leads an evolutionist to conclude that all animals are related. He says that because many living things contain the same chemical substances, modified in ways appropriate to the life of that special individual, all of these living things must have a common ancestor. One author says that many animals have digestive enzymes much alike only because they have similar ancestral backgrounds. He also says that thyroid from a sheep can be used by humans and insulin from a pig is used by man.

Before we give more illustrations, keep in mind what was said earlier: God seems to have used a basic plan in creation and allowed diversity to meet specific needs. Is not this explanation as logical as to say that all things have come from one first single cell?

A commonly given argument from the field of biochemistry concerns serology and blood tests. The classical work tells of human blood serum being injected into a rabbit to cause the production of immune bodies, or antibodies. The serum then taken from the rabbit is called antihuman serum because it will react against human serum to form a white precipitate. In criminal cases this test is used, because only a very small amount of material is needed to determine if it is or is not human blood.

When the precipitation test is used with the blood of different animals, various amounts of the white precipitate are formed. This is said to show the degree of evolutionary relationship between the animals: the more precipitate formed, the more closely, supposedly, the animal is related to man. A chart is prepared to show the result of the amount of the precipitate formed when antihuman serum, prepared in rabbits, is mixed with human blood. Taking the result when the serum is mixed with human blood as 100 percent precipitate, such a table might show: gorilla, 64 percent; orangutan, 42 percent; baboon, 29 percent; dog, 3 percent; and horse, 2 percent. It is then said that the amount of precipitate corresponds closely to the presently accepted degree of kinship between man and the other mammals, decreasing from man to the horse.

As you think about the validity of this argument, do you see how it is based on the premise that similarity shows degree of relationship, giving evidence of evolutionary descent? Review chapter 3 to see that this conclusion is not a valid assumption. As we see similarities in the anatomy, embryology, and functioning of different organisms, we would naturally expect to see the same basic composition in such things as their protoplasm, hormones, and enzymes.

In the field of blood testing, it is interesting to note that there are tests which can be used to show that even human beings are not related to each other! When blood is transfused from one individual to another without proper testing and selection, if the types are antagonistic, fatal results may occur. This can be true even among people of the same family. Because their bloods are incompatible, one can neither assume that they are not related nor that they are not members of the human race. In this case, instead of similarity being evidence of relationship, dissimilarity appears where there is known close relationship.

It is not to his credit that, especially in this field where admitted likenesses are found, the evolutionist seems to have selected for his use that material which shows the point he is trying to make, and disregards the other facts.

In the same way there are other types of blood reactions that point to a close relation between animals that are very different structurally. To make this even more clear, consider a hypothetical situation. Use the five animals listed previously—gorilla, orangutan, baboon, dog, and horse. Literally dozens of substances in their bodies could be chemically analyzed and the percentages of likeness to the same substance in man could be put into table form.

First, use the antihuman serum as already described. Then determine the acidity of the stomach, the salt content of the blood, the percentage of iron in the hemoglobin, the amount of phosphorus in the protoplasm, the exact percentage of calcium in the blood—one could go on and on. As each test is run, tabulate the results, and then study and compare them one with the other.

All the results will not be the same as was found with the antihuman serum. Instead of showing that the gorilla is closest to the human and the horse the farthest away in chemical similarity, the order within the tables will be varied. There might be many combinations of chemical likeness among these subjects. If one wanted to show some special relationship among these five animals, he might present only the tables that substantiated his theory. Selecting only certain facts for presentation seems to have been done in this field of biochemistry.

Summarizing, we see that chemical resemblance is not a good basis for establishing relationships.

First, creation would account for some biochemical likenesses as seen in many plants and animals. For example, most plants contain chlorophyll; all living things contain nucleic acid (most of them the particular one known as DNA used for the vehicle of heredity), and use ATP for energy transfer and RNA to direct protein synthesis.

Second, note that specific amounts of chemicals present and the way they react to each other sometimes give contradictory results. This makes them unsafe guides to use in establishing evolutionary relationships. Unfortunately some of these results found their way into secondary school texts about fifty years ago, and the same ones have stayed there ever since.

Third, consider again that, of over half a million plants and one million animals, God could have made each one unique. The fact that He has used some similarities in basic construction in no way argues for evolution or against a common Designer.

We believe that it is both logical, scientific, and biblical to conclude that God is the Creator and Designer.

8

Why Do We See Plants and Animals Unevenly Distributed?

WHY ARE KANGAROOS found only in and around Australia? Why are llamas only in South America? Why are there no bears in Africa and no elephants in North America? Everywhere one sees an uneven distribution of plants and animals on the earth's surface, but the reason is not easily found. This makes geographic distribution an interesting subject to study and a problem to solve. The answer which we cannot accept is that only evolution explains naturally the strange distribution of some plants and animals on the earth.

Most high school textbooks devote considerable space to this subject and come to the same conclusion. The illustration of the marsupials—pouched animals—is typical. They are supposed to be lower on the evolutionary scale than are the placentals, which nourish their young to an older age before birth. It is said that marsupials are found largely in Australia and Tasmania because these islands separated from the mainland of Asia before the placental mammals had developed. The evolutionist thinks that in this area animals have evolved only as far as the marsupials, whereas on other continents the animals continued to evolve on into the greater variety of nonpouched animals seen today. Evolutionists state that they had long been perplexed by the distribution of plants and animals but that explanations such as this have "solved the puzzle."

As we think about the marsupials, we must realize that the group includes more than the kangaroo. There are also the koala bear, the opossum, the wallaby, the Tasmanian wolf and Tasmanian devil, the wombat, and phalangers. Since the opossum is native to North America, we cannot say that only in the Australian region did animals evolve as far as the marsupials, while on other continents they evolved farther into all of the kinds of placental mammals we see today. Marsupials and their fossils are found on other continents, so to us this does not solve the puzzle. It seems reasonable to suppose that the first mammals were created in different places, as the biblical record indicates, after their kind. Nowhere does the Bible define or set the boundaries and limits of these kinds. One of them could have been the ancestor of today's marsupials.

Unequal geographic distribution is one of those areas to which we referred when we said that it is not the *facts* with which a creationist disagrees, but rather the *interpretation* of these facts. All agree that there are certain continents, islands, or even small areas where some plants or animals are present and very similar continents, islands, and areas where they are conspicuously absent. This must be explained.

Living things usually spread out from a single location along geographic corridors until they find a place with suitable climate, food, and lack of enemies. Eventually they are stopped by barriers such as water, mountains, or deserts. Out of a large group, a few plants or animals by some means will be able to cross a particular barrier and settle down to make it their home. The animals on both sides of the barrier have the same ancestry, but they are now separated from each other. What will happen?

As the two now-isolated groups continue to grow and reproduce, differences will become noticeable. These will arise from mutations, hybrids, recombinations, and translocations of their genes. (More about these changes is written in chapter 15.) Some of these changes will take place on one side of the barrier and maybe not on the other. Gradually these two populations will develop differences—but not ones great enough to account for amoeba-to-man evolution.

Usually the barriers are physical, but sometimes they may be physiological or may appear as differences in behavior patterns. For instance, young salmon from a particular stream are caught and tagged. Later these same salmon, now grown to maturity after having lived in the ocean with salmon from other streams, are recaptured in the same streams where they hatched. Tagged salmon are not found in adjoining streams. Behavior patterns kept them isolated from those in other streams although they came into contact with each other in the ocean. Even though physically alike, these salmon have different physiological behavior patterns, and so they act in different ways. Here again we see change, but not the kind of change that could ever cause evolution to take place, for they are still salmon.

Land bridges are also closely connected with the irregular distribution of living things. Evidence indicates that North and South America were not always connected as we now see them. We think that North America and Asia were once connected through the Bering Strait. The fact that land areas have not always been connected as they are now helps to solve some problems.

Although problems still exist, there are several conclusions which we may reach concerning the irregular distribution of plants and animals on the earth. We have mentioned that such things as isolation, barriers, land bridges, corridors, and migration are connected with these changes. Various combinations of them help to account for some of the changes, thus resulting in uneven distribution. However, the important question is, How much change can be produced by these and other means? In the illustration about salmon, we said that the salmon were still salmon. We could give many similar illustrations including ones about the distribution of rabbits, squirrels, foxes, bears, and dozens of others. Basically we see that, even though there has been some change, the plants and animals, both living and fossils, stay within the limits of each originally created kind.

Just what the created kinds were, we cannot be sure; but we know that at least in some cases, the kinds were not the same as what is meant today by the term *species*. Hundreds of thousands more living things are classified today than when Linnaeus first popularized the word *species* in the 1700s. Scientists are still trying to learn which groups were the originally created ones, and which have developed as the result of the various kinds of change, both the type of changes mentioned in this chapter and the genetic ones explained in chapter 15. By studying the gaps found in the fossil record and by studying the gaps seen in living things, we can be sure that the changes which have taken place are within scripturally indicated boundaries.

9

Should Information About Embryology Concern Me?

EMBRYOLOGY, which traces the early stages in the development of an organism, is a fascinating subject. As a creationist approaches this field, he wonders at the power of a miracle-working God who can cause an egg to develop into the various tissues, organs, and systems of a perfectly functioning individual. On the other hand, a materialistic evolutionist, because he is working within a different framework, tries to find information which will strengthen his theory. Today he is doing this with much less success than he thought he had in the past.

Not too many years ago, all high school science books explained what is called the recapitulation theory. Many young people like the phrase *ontogeny recapitulates phylogeny.* This means that as an individual embryo develops (ontogeny), it passes through the same stages as its remote ancestors did (phylogeny), or racial history. In other words, it claims the development of the individual parallels that of the race. In humans, the changes that take place during the nine months of gestation are thought to recapitulate (repeat) what took millions of years to accomplish by organic evolution, as single-celled animals became the complex ones of today.

Besides explaining this idea, older high school texts usually contained a series of pictures giving this same material in pictorial form. A typical series might show as many as seven animals passing through seven different stages. Each would be first represented by a single cell, then each would be shown as a ball of cells, later by a stage that resembled a fish, still later by one having a tail. Finally the adult animal would be shown.

How should this material be evaluated? First, let us look at what is now being said by the evolutionists. As more has been learned about embryology, he has put less emphasis on the recapitulation theory. About likenesses that are seen in embryos (and they do exist), he seldom says that they actually show a repetition of the supposed history of the race. One author admits that some scientists have recently arrived at the "less romantic view" that these likenesses are just the way things happen to look at this stage of embryonic growth. Another says that the whole idea should be viewed with caution. Most of the authors who still present this part of the evolutionary theory fall back upon the idea that similarity shows relationship.

One declares that "naturally" organisms that are related show resemblances. Another says that the more closely related the animals, the more similar are their embryos.

Most of the high school texts now being used still give at least part of this idea, but few actually say outright that the development of the individual parallels that of the race. Other devices are used. One is simply to ask questions. This allows an author to instill his evolutionary idea into the minds of his readers without actually committing himself to a teaching which seemingly is becoming outdated. This is the same technique used in courtroom trials by lawyers who know that a certain type of question will be stricken from the record, but who also know that the asking of the question will influence the jury's thinking to move in the direction desired. If pictures showing the similarity of embryos are used, students might be asked to think of a reason for the similarity or to observe in what way the early embryos are alike. There may be the suggestion for them to notice which embryos most resemble the human embryos.

Some of the books that use pictures comparing the various embryos and yet making neither interpretations nor conclusions leave these to be made by the reader. Accompanying such pictures, one book makes only two statements: first, study the figures very carefully; and second, trace the development of each of these vertebrates, comparing their development.

Now let us consider the way a creationist might interpret the likenesses seen in embryos. We have said that likenesses do exist, especially in the early stages. The whole setup of the drawings and sketches emphasizes these resemblances. If a person had before him the actual embryos instead of the diagrams, he would be able to see more differences than the pictures show. They are usually drawn with the same curvature and the same size. Only one author includes the information that the drawings are not on the same scale. Several books use the picture referred to above showing seven stages of seven different animals. These are the shark, lungfish, salamander, lizard, chicken, chimpanzee, and man. In each case, the drawing of the adult is shown the same size—about what could be covered with a penny! There is obviously a purpose for picturing the adults—even a lizard, a chicken, and man—as the same size. This purpose, of course, is to show that resemblance indicates ancestral relationship, thus supporting the theory of evolution.

In the 1800s a set of such drawings was made and even with their imperfection, they continue to be published in the texts of today! Weisz says in his book *The Science of Biology*, that these drawings are still being used today "under various guises" even though the author's views are now largely discredited.

We should also consider the fact that many of the resemblances are very superficial. Each embryo invariably develops into the kind of individual that produced the egg. Chicken eggs develop into chickens and frog eggs into frogs. The best way to see how unlike embryos really are is to let them grow, each into its characteristic form. However, actual likenesses do occur and should be explained. For example, blood-vessel patterns appear in some embryos and then are replaced by ones that are seen in the animal at the time of hatching or birth. Rather than looking at these blood vessels as repeating some stage in the supposed past history of the individual, is it not more reasonable to suppose that they serve some useful purpose? As the embryo grows in an orderly way into an increasingly complex animal, changes must naturally be made in both the internal and external parts.

There are only two stages through which the embryo is supposed to have passed that are mentioned in all of the books. These are the one-celled animal and the fishlike creature. A few years ago, many books also included other things such as a hairy creature and a tail stage. These have fallen into disrepute and are now usually omitted. Of the two stages most often mentioned, first think about the one-celled stage. Typical statements are that animals "still" start from a single cell, the simplest form of life in history, and that the fertilized egg theoretically corresponds to a protozoan.

The proposition that human individuals begin life as one-celled protozoans is misleading. They begin life as a fertilized human ovum or egg, which is very different in structure and function from a one-celled animal. This ovum can neither reproduce itself nor carry on many of the life processes that are found in a protozoan. Life starts in the simplest form and then by repeated cell division gradually develops. The fact that this simplest form is one-celled and that a protozoan is one-celled does not mean that one is ancestral to the other. We see no basis for accepting this recapitulation part of the evolutionary theory. It is just a different form of the idea that similarity shows relationship.

In connection with the fish-like stage, the heart and lungs are sometimes mentioned, but emphasis is usually on the gill slits. In his book, *Biology*, Villee remarks that at one stage the human embryo has gill slits. Not too many years ago some books discussed the "gills" of human embryos, but the idea of gills—so obviously not the case—changed to "gill slits." Since actual slits seldom exist, the references now usually are to "gill ridges and pouches." Using the concept of gills is misleading, for these are organs of respiration. In the human embryo, the ridges referred to as the gill ridges become parts of the tongue, lower jaw, and neck of the developing embryo. The depressions between the ridges do not break through into the pharynx, have nothing to do with respiration, and so are not comparable to the gill slits of a fish.

Very few of the present high school authors still make any reference to the tail or to a hairy stage.

In his book *Biology*, Kimball says that the human embryo has "temporary possession of a tail." What is referred to, instead of being a tail, is the coccyx or terminal portion of the backbone. It is true that at one time in the developing embryo the coccyx projects beyond the surrounding parts, because it has a faster rate of growth. However, that never makes it anything other than the end of the backbone. A seldom found, but chance exception, to this is mentioned in chapter 11.

In the whole field of embryology, there are many unanswered questions: If the developing embryo is supposed to reenact stages in the history of the race, why are so few stages included? How can such stages as the larva and pupa of an insect be explained? Why are recapitulation stages seen in only a few parts of the embryo, whereas other parts do not show them? Why are the changes in a developing plant embryo considered to be of little significance? In fact, there are so many unanswerable questions and problems that most authors say the evidence from this field should be viewed with caution. After pointing out some objections, one author, with whom we agree, says that recapitulation simply does not occur.

There is no doubt that God could have used a different pattern for each developing organism. He could have made hundreds of thousands of different ways for a single cell to develop into an adult so that each plant and animal would have been entirely different from any other. However, this does not seem to be the case; and in their development, as well as in their adult form, God seems to have been consistently conservative. Since a general plan of growth was suited to the individual and to its environment, it would appear to be true that He used it over and over again. Thus, rather than accept the recapitulation theory, we look at temporary embryonic structures as either having some function in the developing embryo or being part of the general creative plan. Each part was modified to fit the needs of that particular individual.

10

Is It Optional What We Think About the Source of Design?

THERE IS STILL ANOTHER unanswered question concerning the theory of evolution, which is larger than most of the others. This is the problem of design. When we examine the eye of an insect and that of a man, we are amazed at the intricacies of each and how they differ. In an entire lifetime of study, the human mind cannot take in all there is to know about the functioning of a single part of the human body. We continue to wonder at daily discoveries in organ transplanting, complex enzyme interaction, atomic structure, precision spacecraft engineering, and a variety of other scientific fields.

To think that the order and design we see in the world around us could possibly have happened by chance or accident takes much more faith than to know that God planned things this way. Design must have a designer. The idea of an all-powerful God cannot be ruled out by assumption and guesswork. The theory of evolution is as shallow as saying that a watch, a great painting, a skyscraper, or the computations necessary for an interplanetary flight just happened with no mind to plan them.

It is more old-fashioned and unreasonable to believe that by evolution our universe could accidentally have come into existence and accidentally continue to be maintained, than to believe that God created and sustains it.

11

Why Must I Make a Decision About Useless Structures?

VESTIGIAL STRUCTURES are parts of the body that are usually considered useless. An evolutionist thinks of them as remnants from a useful state in some ancestral animal. One author mentions man's appendix, his scalp and ear muscles that can be contracted, and the fused bones at the lower end of the spinal column. He says their presence brings people to the conclusion that man descended from ancestors which needed and used these structures. One author points to the fact that whales have no hind legs but a number of small bones on each side of the spinal column in the regions where one would expect hind legs to be attached. He also mentions that snakes and porpoises have remnants of hind limbs. He comes to the final conclusion that the explanation for these "useless structures" is that the animals are descended from a common ancestor. These structures can be considered "remnants" of structures once useful to the ancestor.

A creationist combats this line of reasoning with the fact that in recent years the list of so-called vestigial structures has been greatly reduced in number. The high school text written by Gregory and Goldman with a copyright date as recent as 1965 says that man has more than two hundred vestigial structures! Most books mention less than half a dozen. Formerly even the thyroid and pituitary glands were listed. These are now known to be essential glands. The pituitary is of such prime importance that it is frequently called the "master gland."

It has been said that our list of useless organs decreases as our store of knowledge increases. As a part is studied and its function discovered, it is often removed from the list of useless organs. We should hesitate to call any organ useless, because most of the former so-called vestigial ones have been found to have some function.

We should realize, too, that the degree of usefulness is not the point in question. Different parts of the body have different degrees of usefulness; but if they have any usefulness at all, they cannot truly be called vestigial. Because thousands of people have suffered no harmful effects from appendectomies is no argument that an appendix is vestigial. In the same way, an arm or leg may be amputated with little harm to the body, but this is no reason to call the arm or leg vestigial. Some scientists think that the appendix may not be useless, even though they do not agree on the amount of usefulness, or even on what that usefulness may be. Some think it adds lubricating fluids to the contents of the intestines; others think that it may have an endocrine function; others say it secretes small amounts of digestive juices; and still others think it may manufacture some blood cells and functions in protecting the body from diseases. It may possibly have a combination of these functions, but we are quite sure that it is not *vestigial* in the true sense of the word.

What we have just said about the appendix may also apply to other so-called human vestiges; but we will say an extra word about the "tail" because it involves a slightly different explanation. Kimball says that the fused vertebrae which make up the base of the human spine could be vestigial remnants of a tail posessed by our ancestors. He points out that human babies occasionally are born with short tails (which of course are removed). We do not agree with Kimball for three reasons.

First, we cannot entirely discount the fact that human babies are sometimes born with a short projection which looks somewhat like a tail. However, this could be due to a deranged process taking place while the embryo is developing. When the normal growth is altered, we sometimes see the result in the form of Siamese twins, cleft palates, or harelips. No one would say that these were once normal conditions in a remote ancestor. A "tail" might be such an anomaly.

Second, most scientists agree that there is a time

during embryonic growth when the coccyx grows faster than the surrounding parts. This different growth rate causes the coccyx to project beyond these adjacent areas, but this projection of the coccyx has nothing to do with its being vestigial.

Finally, the coccyx, or end of the backbone, is not really useless! Some very useful muscles are attached to it, so it is not truly vestigial.

We sometimes say that God was conservative in the use of His creative patterns. This fact may explain the conditions we find such as in the hipbones of a whale. These are cartilage-like structures embedded in the place where hind legs would be if a whale had hind legs. The whale is a mammal possessing hair and milk glands. Since it is a mammal, it is reasonable to suppose that God could have created it on the same general plan that He used for other mammals, making modifications to fit it for life in the water. In the same way the bat, another mammal, was created on the same general plan but with wings and other modifications to fit it for flying.

When thinking about other so-called vestigial structures, consider the following possibilities:

1. Could it be the result of God's conservative way of creating a number of animals built on the same general plan, but each modified to fit a special environment?

2. Could it have some use, maybe slight or as yet undiscovered?

3. Could it be the result of some abnormality taking place while the embryo is developing?

4. Could it be the result of degeneration from the way in which God first made that plant or animal?

We may not be sure which of the things mentioned, or what combination of them, accounts for so-called vestigial structures. However, we can be sure that any of the explanations, when properly interpreted, will be in harmony with the truth of God's Word.

12

How Does the Origin of Matter and the Universe Concern Me?

WHATEVER ONE'S BELIEF is concerning these important subjects, an element of faith is involved. Most high school texts make no mention of God or of the supernatural. If a person has ruled these two beliefs from his thinking, he must come to the conclusion of one of our texts which says that there is no real knowledge of how the sun and its planets came into existence. Another book says only that organisms are made from the same materials that make up the rest of the world. No mention is made as to what the source of these materials might have been.

Organic evolution is usually spoken of as being naturalistic or materialistic. The dictionary defines *materialism* as any theory which considers the facts of the universe to be sufficiently explained by the existence and nature of matter. *Naturalism* is defined as the doctrine denying that anything in reality has a supernatural significance. Since the authors of most high school texts are working within the framework of these definitions, do you see why they have no real explanation for the origin of matter or of the universe?

Admittedly, Christians approach this subject with a preconceived idea, for miracles are tied in with every part of their faith. If miracles are removed from the Bible, there is left only a useless shell. It is not logical for a Christian to exclude miracles from his thinking, even though he will never be able to comprehend much about the original creation. When asked about the origin of matter and the universe, a Christian says, "Faith enables us to perceive that the universe was created at the bidding of God—so that we know that what we see was not made out of visible things" (Hebrews 11:3, Twentieth Century New Testament).

Faith is the basic requirement for understanding the universe, with all the material from which it is made, as well as all that it contains. To summarize, Christians accept by faith the Bible record which teaches that for His glory and according to His own sovereign will, the triune God created; He created both visible and invisible; He created all things without the use of preexisting material. If He were not able to do this, He would not be God!

Some years ago Christians were laughed at for believing that all matter in the universe was made out of "nothing." In recent years, some of that faith has been turned into sight. We continue to learn more about the structure of atoms and of the electrons, protons, and neutrons of which the atoms are made—particles of matter so small that billions and billions of them are found in a single drop of water. We now realize that electrons and protons have electric charges which are, respectively, negative and positive. The Christian is not surprised to hear that energy and power have always been tied up in matter, for God made the earth by His power (see Jeremiah 51:15). While energy and power have always been in matter, it has been only recently that man has learned to release tremendous amounts of it by atomic fission.

Since science is not static and our world is continually changing, as we continue to study it, faith is often changed into sight. Today's theories may have to be changed tomorrow as new observations are made and new experiments are carried out. The two-way communication instrument of Dick Tracy and the space travels of Buck Rogers some years ago seemed improbable and fictitious, but now they are possibilities.

For many years man has been able to change matter into energy and energy into matter. However, the fact that they are sometimes interchangeable does not help the unbeliever explain the origin of either. The inability to do this is one of the great weaknesses of the theory of evolution. Before matter and the universe originated, some eternally existing source of power had to exist. The Christian recognizes this when he says, "God has made the earth by His power, He has established the world by His wisdom, and by His understanding and skill has stretched out the heavens" (Jeremiah 10:12, Amplified Bible).

13

Is Knowledge About the Origin of Life Important to Me?

EVERYONE HAS PROBABLY WONDERED about the origin of life: What is life? How did life begin? What details are connected with the statements "God created the beasts of the earth" and "God made man in his image"? It is especially interesting that theories about life relate to our discussion of evolution.

It is possible to skip over all of these questions in a rather meaningless way. One textbook author tries to explain life's origin by saying that through a method not known, giant molecules became possessed with the power to produce others like themselves, starting life in its simplest form. This statement really tells us nothing, unless it might be that the author probably does not believe in miracles!

Most high school books say little about the origin of life, unless to suggest that this subject should be left to the philosopher. However, if a person's theory of development is reasonable, it will not preclude a reasonable explanation concerning the *beginning* of that development. This, of course, brings us back to the subject of miracles and of a personal God. We cannot agree with the books that say this question is outside of the realm of science.

In all the books we are considering that do mention the origin of life, the following four possibilities are mentioned: (1) special creation; (2) life has always existed on the earth; (3) life came here from another planet; or (4) life originated from nonliving material.

The first one is ruled out by many scientists. Many will say that living things originated from physical and chemical properties of the ancient earth and that nothing supernatural was involved.

The second idea, that life has always existed, is so seldom considered that it is included here only for its historical interest.

Some books seriously consider the third theory, which suggests that by means of something like resistant spores, life could have come here from some other planet. If this were true, it would not be life as we know it, for it would have to stand tremendous extremes of heat and cold and deadly radiation. Then, too, if life arrived here from some other planet, there still remains the question of how that life had previously begun. That answer only pushes our problem back one step farther.

In many books one may read historical accounts about the fourth idea, which is called spontaneous generation. This idea suggests that life originates from nonliving material. These accounts tell especially of the work of the famous Italian scientist Redi and of Louis Pasteur. People used to think that meat produced worms as it decayed, not realizing that tiny eggs in the meat had been laid by flies, and that these eggs hatched into maggots. Redi excluded flies from meat to show that under these conditions maggots were not formed.

In the middle of the nineteenth century, Louis Pasteur performed many experiments to show that nonliving material was easily contaminated by living organisms such as bacteria, virus, molds, or yeast. These are usually found in the air, and Pasteur repeatedly demonstrated that food and other material would not produce living organisms after they had been sterilized properly. In fact, today one may see flasks in the Pasteur Institute in Paris, unchanged since Pasteur sterilized and sealed them before his death in 1895.

As a result of the work by Pasteur and others, we accept today the principle that life comes from life. About spontaneous generation (or *abiogenesis*), the second edition of *Webster's New Twentieth Century Dictionary* says, "a former theory, now abandoned." The disproof of spontaneous generation leaves the scientists in a dilemma. Look back at the four possible explanations for the origin of life. If a person does not choose to believe in a God who can create life, and if it is unscientific to believe in spontaneous generation, what then can he do? The unbeliever is forced to make a decision, and he usually decides in favor of spontaneous generation! He knows that the world was once without life, and that life later appeared; and he knows that this must be explained. You rightly ask, How can a scientist possibly believe what is considered unscientific?

In a few years it will likely be superseded, but now it is said that the way out of the dilemma about the origin of life is to accept the work of men such as Oparin, Urey, Miller, and Fox. They concede that as a rule life does not start spontaneously now, but it might have in the past done so. Some of their experiments involve methane, ammonia, water vapor, and hydrogen. These four substances, supposedly present on the primitive earth, were put into an airtight apparatus. They were then circulated for one week past a high energy spark, while heat and water vapor were supplied from the outside. When at the end of the week, an analysis was made, complex molecules not present at the beginning were found. To be living material, these molecules would have to be able to reproduce themselves; and it would not be surprising if some day this were done in a laboratory. If life is ever produced in a laboratory, would this be considered as evidence for evolution? Would it not rather show that a very, very superior mind working with the best equipment, is necessary to do this? It could not have happened by the laws of chance.

The average cell is so complicated that for it to

have happened by chance is a statistical monstrosity! Cells differ in their complexity. To present an idea of how complicated one is, consider a little of what the average high school student is expected to know about the cell's composition. Within most cells is a nucleus. Within each nucleus are chromosomes. Within the chromosomes are genes, and we now think that each gene is made up of one (or part of one) molecule. This is a deoxyribonucleic acid molecule, commonly called DNA. Each molecule is pictured as a double helix by the scientists Watson and Crick. This ladder-like spiral is made up of alternating deoxyribose and phosphate portions of nucleotides. These are linked across from one side of the spiral to the other, like rungs on a ladder or treads on a staircase, by purines and pyrimidines, of which there are four kinds—adenine, thymine, guanine, and cytosine. Because the helix is often so long and the various parts we have listed can be combined in different sequences, the number of kinds of DNA molecules is almost unlimited. Someone has said that the possible kinds of combinations within one long DNA molecule are as many as are the number of atoms in our earth. Much direct evidence has been learned about DNA by laboratory studies of simple substances such as bacteria, viruses, and mold. This complex DNA material is now thought to transfer genetic information from one generation to another and to help control cell activity and development. At the time of cell division these molecules are capable of duplicating themselves, so that each of the billions of cells in our body has exactly the same kind of DNA molecules. (See chapter 16.)

Would it not take a lot of faith to believe that such a fantastically complicated molecule could come into existence merely by chance? Then too, the average cell has many, many more complicated structures. One could write a book as thick as an unabridged dictionary and still not include nearly all the facts known about cell structure and function.

It is beyond our mental comprehension to think that by chance the parts making up the structure of cells could simply happen. It is also beyond comprehension to think that by chance, all at the same time, such necessary life processes as digestion, respiration, excretion, and reproduction should happen to come into being.

It is reasonable and scientific, as well as biblical, to believe that an all-powerful God is the Author of both physical and spiritual life. He originated and sustains life. Each day we see the miracles evidenced by these omnipotent acts.

14

What Should We Know About the Genesis Kind?

THE QUESTION ASKED by the title of this chapter makes us think of other similar ones. Did God create each of the more than two hundred kinds of domesticated dogs? Did He create the Great Danes, the greyhounds, the collies, the shepherds, the bulldogs, the poodles, and the Pekingese, each one as we see it today? Evidence does not point to this. In the stories we read about the Middle Ages, it would seem that only a very few kinds of fierce, wild dogs roamed the countryside and the cities. Were these dogs ancestral to all of the present-day varieties? It would seem so; and new kinds of dogs are still being produced today.

Realizing that it is possible that all dogs may have come from one ancestral type, let us think back one step farther. Consider whether God might have created an animal kind ancestral to not only dogs but also wolves, jackals, and coyotes. This is probably true. The evidence we see in nature and in the Bible tells of His creating large groups, such as fish, birds, cattle, and beasts of the field. Since so little detail is given in the Bible, we cannot be dogmatic, but it is possible that the situation may be similar to what is pictured in the sketch on page 82.

If this sketch actually represents what took place—and is still happening—considerable change may have taken place within each created kind. Some of these changes have taken place naturally, and some can be produced artificially. All Christians, maybe without realizing it, accept change on some of these levels. This must be done when one considers that all human beings descended from Adam and Eve. More will be said in the next two chapters about how these changes could take place; therefore, the three chapters should really be read as a single unit.

The Bible tells us that God created things "after their kind." It does not go into a detailed and scientific account of the kinds. One could make a hodgepodge list of scientific and nonscientific terms such as *species, group, race, genus, individual, family, order, class, subspecies, type, breed,* and *variety.* From such a list no person has license to pick one term, and only one, and say that this is the one kind that God created. The Bible does not tell us that God created all the different races or all the different species or all the different families. Certainly the evidence is against His having separately created such individual kinds of cow as the

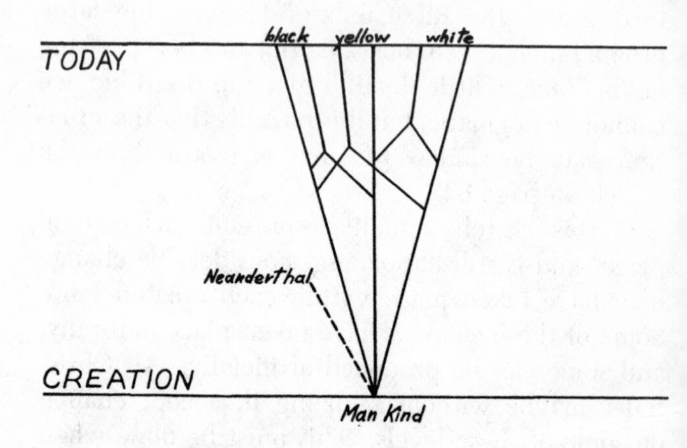

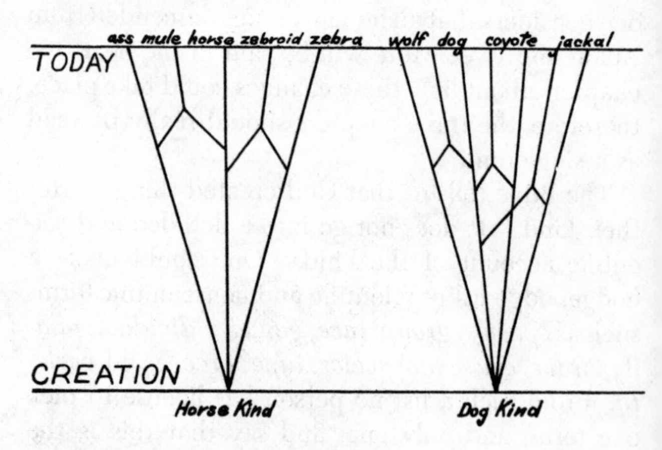

Figure 2

Chart showing the relationship between a possible created "kind" and forms living today (from Wayne Frair and P. William Davis, *The Case for Creation*).

Jersey, Hereford, Aberdeen Angus, Holstein, Guernsey, Brown Swiss, and Shorthorn. It appears that He created a type ancestor from which, by limited change, have developed all these different kinds of cows.

The same reasoning is thought to be true concerning classification groups. Many Christian scientists think that evidence points to God having created an ancestral type of jungle hen. From it, by laws He put into operation, have come the Plymouth Rocks, Bantams, Rhode Island Reds, and other chickens. Changes such as this are acceptable within the lower classification groups. This in no way commits one to accept the kind of change among higher groups that would have been necessary for organic evolution.

One must differentiate between the amount of change that is acceptable to the creationist, and the amount that is not acceptable. Of course, the ground for judging this difference would be the Bible, if it touched on each point–telling about each plant and animal. However, it does not go into specific detail in this area. The Bible says only that God created things "after their kind" or, as some translate it, "according to their subdivisions." What were these subdivisions, or kinds? As indicated, they were not always what is designated by the word *species* or any one term in our man-made system of classification.

It is unfortunate that there are Christians who hold to the term *fixity of species,* which is neither biblical nor scientific nor logical. Part of the reason so many people hold to this idea is that they have little knowledge of what is meant by a species. We realize that species is a term which cannot be rigidly defined, but the following explanation may be helpful. Members of a species are usually considered to be plants or animals similar in structure and function, that do not customarily breed other than with their own members to produce offspring that are fertile. Members of different species are individuals that do not usually cross successfully with each other to produce vigorous offspring. We have said usually, for some organisms will not interbreed in natural conditions but will if placed in artificial conditions. This is true of mallard and pintail ducks.

The average untrained individual who says, "God had to create every species," often is confused by plants and animals of the same species that look different because of sex, period of life, and location. Members of the same species have similar genetic content, but sometimes this cannot be told by looking at external characteristics. The two sexes of many animals, like the turkey and the peacock, look very different yet compose the same species. There are also animals that at different times in their life history have great variations in appearance. Unless a person knew something about the life cycle of a butterfly, it would be hard to tell that an adult belonged to the same species as its larva and its pupa. The same species may look different at different times of the year, as are reddish-brown weasels in Northern regions, called ermines because they turn white in the winter. The reverse may also cause confusion: plants and animals which look very much alike can be members of entirely different species.

Let's look at a few more illustrations to see why one cannot take a particular species and say, "This is what God created, and it remains as this particular species for all time." There are common leopard frogs that range from Vermont to Texas. This is clearly a frog of the same species, and all along this geographical line they freely breed with each other. However, when the frogs from Vermont are bred with ones from Texas, the offspring are abnormal. There is enough genetic difference between these two that we would classify them as different species if we did not know that they are the same kind of frogs but with a scattered geographical distribution. Such a case as this frog gives evidence that a species is neither a uniform nor unchanging group of organisms. This concept of change is built into the definition of a species that says, "Species are a Mendelian population which live in a more or less extensive area and which gradually alters its genetic composition." Such change can also be seen through the work of Luther Burbank. By crossing the plum and apricot he produced the plumcot; a daisy from New England and one from Japan were crossed to get the beautiful Shasta daisy. The loganberry is a result of artificially crossing a blackberry and a raspberry. God surely did not create a plumcot, a loganberry, or a Shasta daisy.

The mule is another good example to show that the species is not a created and fixed entity. A hybrid of a female horse (mare) and a male donkey (jackass) produce this animal with desirable work qualities. Every once in a while a fertile mule is produced, even though the parents belong to separate species. A hinny, the result of crossbreeding a male horse (stallion) with a female donkey (jenny), is not so useful an animal as the mule. However, neither the mule nor the hinny is a created species. One can readily see why the term *fixity of species* should not be used. This term has existed since the classification work done by Linnaeus in the middle of the eighteenth century. It is used by people who mistakenly think the term *species* used by Linnaeus is synonymous with the kind created by God as recorded in Genesis. They do not know that two species can be crossed to form new ones, and that sometimes what was once thought of as one species should be divided into two or more.

What are the "kinds" of Genesis? We do not know, for the Bible does not say. It shows that God created groups of living things. He surely did not create so many that it was beyond Adam to name *all*, for we read that "whatsoever Adam called *every* living creature, that was the name thereof" (Genesis 2:19). This one-man naming process is not possible with all the species, for it would take many years to examine each of the species of plants and animals in existence today, even if one spent only a few minutes on each and worked day and night. First Corinthians 15:39 (New Scofield Bible) says: "All flesh is not the same flesh, but there is one kind of flesh of men, another flesh of beasts, another of fish, and another of birds." Genesis lists a few more animals than these four groups. Nature indicates that God created many more major groups than the few listed in Genesis. This is where nature (science) and the Bible must work together. We see around us certain basic groupings or categories. These naturally occurring groups are not always the same as any one division of our man-made classification scheme. In some cases these naturally occurring groups seem to be families, sometimes genera, and in other cases, species. Some scientists would even include some classes and orders. If these are the biblical kind, they surely are not all listed in the Scriptures. One has to turn to God's revelation in nature to learn more about them. These groups can be recognized by the unbridged gaps that exist on each side of them —spaces unfilled by missing links. The existence of hundreds of these gaps does not give us a picture of amoeba-to-man evolution.

There is much study yet to be done to discover what are the kinds that God created. Some people are not interested in this branch of science, but we have said that if you are, you may be one to discover some of the as-yet-unknown facts. Those who work with living and fossil plants and animals often have to make changes in their classification. Do not be surprised at this, for that is how a scientist must work—learning, discarding, changing, and trying again. This is the way men like Edison and Ford had to work. If they had not, we would not now be enjoying the things which they invented. Paul Ehrlich is a good example of a man who worked this way in trying to find a drug to kill the germ that causes syphilis. With 605 drugs he experienced defeat, but it was during his 606th experiment that he found a compound that would cure most cases of syphilis. Although it is now seldom used since other drugs have replaced it, it saved many lives and is still called "606."

Open-minded scientists, Christian and nonchristian, are endeavoring by continual study to determine where gaps between the various groups of plants and animals occur. When we are able to do this, we will know which groups are the created kinds of Genesis. As scientific creationists, we cannot subscribe either to fixity of species on the one hand or to evolution on the other. After examining the question of species and kind, if one firmly holds to the scientific evidence, faith in the authenticity of the Scriptures will also be reaffirmed.

15

Why and How Much Do We See Things Change?

We see plants and animals changing continually. This brings us to some crucial questions. Why do living things change? How much can they change? How much should this change influence one's belief in the theory of evolution or in the Bible? How much information about change is fact, and how much is interpretation?

To find answers to many questions, we must turn to the field of genetics, which is the science that deals with heredity. This is a rapidly changing field. As we look at a few things being done in the field of genetics, it is with the realization that by the time this book is published, such strides may have been taken in this field that these things will seem very out-of-date! For instance, sexual reproduction is sometimes bypassed so that coconut milk will cause the female cell of a carrot to start developing. Frogs have been produced by removing the nucleus from an unfertilized egg and replacing it with the nucleus from an intestinal cell of a tadpole. The resulting individual looks like the tadpole rather than like the frog which produced the egg. Before birth, rats can be treated with pituitary extract. By this method much has been learned about why some rats are bright and some stupid.

The results of these studies and experiments have produced a whole new science called fetology. Many babies have been saved by giving the fetus blood transfusions. Cataracts and cirrhosis of the liver may develop in children with the genetic disease of galactosemia. If detected early enough, adverse effects of this disease may be prevented. This is now being done.

Living things do change, and since the evolutionist says that they change enough to cause everything to be related, we should look at some historical background. We cannot here go into the ancient history of changes which have been produced, but there were men even before the birth of Christ who had various ideas along this line. Before we consider several scientists who proposed methods of change, we should mention the German philosopher, Georg Hegel. His work prepared the way for the evolutionary idea that things in general are progressing upward. He said that even history showed progress from lower to higher manifestations in various principles and activities. This surely is not what we are seeing today with the increase in crime, war, delinquency, drug abuse, and vandalism.

Lamarck, a French naturalist working in the late 1700s and early 1800s, introduced many of the terms and ideas connected with the theory of evolution. His theory was based on the idea that organisms passed to their offspring characteristics which had been developed because of need created by the environment of that parent organism. This was known as inheritance of acquired characteristics or the "use and disuse" theory. He thought that the descendants of a blacksmith would gradually develop larger and stronger right arm muscles. Lamarck maintained that giraffes living in arid areas where grass was insufficient had to eat the leaves of trees. As they reached for higher leaves, their necks grew longer, and so did those of the following generations. He said that if apes came down from trees and used their hind legs for running, we would see the gradual development of legs adapted for running and of arms shortened to be used for other purposes. It was a very attractive theory for those who wished to believe in evolution. Even today in Communist circles there are those who hope that by controlling the environment of infants and children, succeeding generations will eventually produce supermen. This was also the idea behind the thought of Hitler and the principles used by the Nazis. The theory of Lamarck is fascinating except that it is not true! All characteristics that heredity passes on to the next generation must be via the germ cells. What is done to body cells is not inherited—and in most cases this is fortunate. Many high school texts tell of the work of August Weismann, who cut off the tails of twenty generations of mice. The tails of the mice of the twenty-first generation were just as long as those of the first generation! One ency-clopedia says that the theory of the inheritance of acquired characteristics is no longer accepted by Western scientists. There is no inheritance of ac-quired characteristics. If you should lose a leg or learn to be an expert magician or have a trans-planted heart, your child will in all probability be born with two legs, will have to learn magicians tricks from the beginning, and will have a heart of his own.

Charles Darwin's name is likely the first to come to mind when the subject of evolution is mentioned for he is the man who first popularized the theory He taught that living things varied in many direc-tions, and that eventually through "the struggle for existence" and "survival of the fittest," new kinds of organisms would evolve. In life's struggle to live, he thought that the weak die and the strong live, with the total pattern leading to stronger and better plants and animals. His explanation for the long neck of a giraffe would be different from that of Lamarck. Darwin said that the length of necks varied in any population. The long-necked ones in their struggle to exist were more successful in get-ting food from the higher branches. This caused the fittest to survive and the long neck would be passed on to the next generation. Thus the unfavorable trait (in this case, the short neck) would disappear.

Scientists agree that selection plays an important part in the changes found in living things. Selection alone cannot initiate changes; therefore, it cannot be responsible for an amoeba-to-man evolution. Especially when combined with crossbreeding or making use of mutations and other genetic changes, artificial selection has helped man develop many useful plants and animals. By choosing to cross cows that produce a large quantity of milk containing high butterfat content with ones having good beef qualities, man can get the kind of animal he wants. With dogs he may select and breed those with either speed and endurance, or beauty and intelligence. Largely by selection, turkeys have been developed to secure birds that suit the needs of small families. By the same method, for use in laboratory work, small pigs rather than large ones have been produced.

Before we draw our final conclusion about the weaknesses of Darwin's theory, we should consider the work of Gregor Mendel and Hugo De Vries. Even though much of Mendel's work was done at the same time as that of Darwin, it received little recognition until the turn of the twentieth century. If Darwin had been aware of its significance, he might have changed some of his conclusions. Mendel was an Austrian monk who, with great patience, systematically crossbred garden peas and kept records for a number of generations. Even though he knew nothing of the existence of chromosomes and genes, on the basis of his experiments he formulated some principles of heredity to which we still hold. One of the chief ways in which Mendel's work was important was that it put a serious question mark at the end of all Darwin's work. Mendel, as did Darwin, recognized that organisms vary. However, he found that they did not vary in all directions, but rather only within certain limits. These limits keep one organism from becoming another and so make evolution impossible. In this connection be sure to read what the next chapter says about the limits to color changes in the eyes of fruit flies.

Mutation is the key word always connected with the work of De Vries. Sometimes an individual very different from others in a population will suddenly appear and is popularly called a sport! Some of the best-known mutations are seedless oranges and grapes, the short-legged Ancon sheep, hornless cattle, albino animals, and plants without chlorophyll. The largest number of mutations are harmful monstrosities. Many of them, like Siamese twins, six-legged or two-headed domestic animals, die a short time after birth.

Now that we know that chromosomes are DNA molecules (or parts of them), by using this term let us try to explain mutations. During millions of ordinary cell divisions, the DNA molecule duplicates itself to produce new cells like itself. Occasionally an unpredictable happening takes place

A different nucleotide is put into the molecule or one is left out. This change in the molecule modifies it so that it is no longer an exact copy of the genetic message to other cells; and so the individuals of the next generation are different from those of the last. It is this change in the genetic message that results in what we call a mutation.

We learn much about mutations by artificially producing them in the laboratory. The most common technique is to treat the reproductive cells with such external agents as heat, chemicals, or irradiation. If this is properly done, one of several things may happen within the nucleus of the cell. (1) There may be a change in the number of chromosomes so that a set is doubled, tripled, and so on. This condition is called *polyploidy*. (2) There may be a change in the chromosomes themselves, as when a piece breaks loose and then fuses onto another chromosome, or fuses with the original one but in an inverted position. These are called *crossing over* and *inversion*. (3) There may be the accidental loss or addition of a whole chromosome. (4) There may be a change in the chemical nature of a chromosome. These four kinds of changes are called *chromosome mutations,* but there are also *point mutations* or changes in a single gene. Both of these kinds of changes breed true to type, and so they are passed on to the next generation.

As we begin to think about mutations being a cause of evolution, we note that most mutations are harmful to the organism involved. It is true that man often makes use of them, but this is an unnatural situation. If the Ancon sheep had not been protected by man, its short legs would have caused it to fall prey to its enemies because it was handicapped in running and jumping. Hornless cattle also lost one of their best means of protecting themselves against their enemies. If not continually grafted by man, seedless oranges and grapes would cease to exist.

Besides usually being harmful, we recognize that naturally occurring mutations are very uncommon. This is especially true of ones that in any way might be helpful; thus it is hard to see how they could contribute anything to the evolutionary process.

Mutations are rare, occur at random, are very unpredictable, and seldom survive. Therefore, since to be helpful in an evolutionary process they would have to follow a fixed direction, they fail in this requisite.

A final reason why mutations are not likely to be helpful in an evolutionary process is the fact that they are usually recessive. This means that they cannot be seen in an individual unless two genes are present, one from each parent. There are two ways that an individual might get two genes for a recessive mutation so that it could be seen in the individual. The first and unlikely one is that each parent received the mutant gene at the same time. A more likely possibility is that the parents each received the recessive gene from a common ancestor. This is the reason that some states have laws against certain close relatives marrying each other. It is thought that Queen Victoria had hemophilia, or bleeders' disease. In order to maintain a royal line, there were more than the usual number of family intermarriages. This led to the premature death of many of the queen's descendants.

Three theories, proposed in the past to account for the way evolution might have taken place, have been examined. Lamarck's is not acceptable because acquired characteristics are not inherited. Darwin's theory of selection, either natural or artificial, accounts for considerable change, but only as it works on changes originated by some other method. Also, these changes are only between specific barriers. De Vries' mutation theory also explains some change, but mutations are rather rare, usually harmful and recessive, and take place in random directions—again, only within the groups that we believe are the kinds spoken of in Genesis.

As you see, none of these three methods will account for evolution. Most scientists will agree with this, but say that some combination of them will be effective. The most frequently combined ones are natural selection working on mutations. We admit that any one method or any combination of methods will produce some change, but it is not the kind of change that would have bridged the gaps between large category groups as necessitated by the theory of evolution.

Recall other things which have been mentioned that will cause some change. Migration was one of these. If an area has long and short-haired dogs and either of these should leave, the population remaining would differ from the original. Associated with this migration, we often find isolation, which in time changes the gene pool. (A gene pool is all the genes contained in a local population of interbreeding organisms. So of course, the gene pool would determine all the hereditary possibilities of that organism at that locality.) However, the dogs are still dogs. Hybrids have also been mentioned, and there are many both desirable and undesirable ones. Some of them may produce new genera, but never organisms of different higher classification groups. God set into operation various ways in which the chromosome numbers may be increased or the chromosomes rearranged, thus giving various kinds of diversification, but none of these should in any way be confused with evolution.

If there is no method by which evolution could have taken place, we need not really even consider further validity of the theory.

16

What Can I Conclude About the Complicated Chromosomes?

In a previous chapter we are told a little about the intricacies of chromosomes and the genes of which they are made. It is no wonder some people feel that they are wandering in a maze when they try to understand the parts of a chromosome or DNA molecule. In describing such a molecule one must use these terms: *thymine, guanine, adenine, cytosine, purines, pyrimidines, deoxyribonucleic acid, nucleotides,* and *phosphates.* The chromosome is surely complicated!

An evolutionist usually says that the evolutionary process can only remodel and build upon what already exists. To be logical, a person who believes in amoeba-to-man evolution must believe that all of the genetic material, all the DNA which will eventually produce all the different kinds of plants and animals in the world, must have been present in the original parent cell. If that cell were produced by spontaneous generation, how could it happen that all of the substances mentioned above could have come together by chance, in the right kinds, proportions, ways of fitting together, and chemical structuring? Using the scientific equation, probability for this chance structuring to have happened without being directed by an omniscient and omnipotent hand would run into astronomical numbers. If, as most evolutionists say, no chromatin material has been added to the original first cell, it would have had to contain all the DNA material possible for making all of the plants and animals of all future generations. Should it not have contained all the material which, if combined in the right way, could produce all the kinds of arms, legs, wings, fins, and different eye-colors of all creatures?

An evolutionist will not admit the truth of the above statements. Weisz says that it would have been useful for terrestrial plants to grow legs, but this could not occur because, supposedly, the ancestors did not possess the necessary structural and functional potential. If the potential to produce legs in thousands of kinds of animals was present in that original cell, why could not legs also have been produced in plants?

Drosophila melanagaster is a small fruit fly commonly used in laboratory experiments. By various methods, it is possible to secure from the common red eyes such colors as claret, white, garnet, vermillion, cardinal, maroon, pink, scarlet, sepia, carnation, purple, cinnebar, brown, ruby, and raspberry. As far as we know, green or blue eyes have never and are not likely to be produced. The answer to why this is true is simple. DNA molecules with the sequence of bases capable of producing these colors do not exist in fruit flies. They should have been there if that original cell contained all of the material necessary to produce everything in all future generations of plants and animals. If material was there to produce green and blue eyes in other animals, why not in fruit flies?

It would seem that the Christian explanation for the lack of green and blue eyes in fruit flies is more logical, for as God created each kind he put within it certain potentials. We do not know how many or what were the kind, but evidently the ones which were the ancestors of the fruit flies did not contain the ability to produce blue or green eyes. This potential was given to some other groups of animals. God did not give to the various kinds of plants the DNA necessary for producing legs; but He did give it to many of the groups of animals, so that is where legs are found.

Now consider the fact that an evolutionist says, "Animals vary in complexity from amoeba to man. Plants vary in complexity from single-celled algae to multicelled flowering plants." If it were true that living things had evolved in this order, would it not also be reasonable to expect to see a gradual increase in genetic material as plants and animals become more complex? Should not this gradual increase in complexity show an increase in the number of genes and also in the number of chromosomes that contain them? Is it not logical to suppose that complex organisms would have more chromosomes than simple ones? Although a difference in complexity and size may sometimes exist, this is not an essential element to be considered.

Selected species of the following plants and animals are arranged approximately in order as they are said to have evolved:

Plant Names	Number of Chromosomes Per Body Cell
an algae	48
another algae	24
a moss	40
a fern	28
pine	24
onion	16
a lily	48
trillium	24
radish	9
red clover	14
alsike clover	8
tobacco	96
peas	14
Animal Names	
radiolarian	1,600
certain roundworms	2
earthworm	36
snail	200, 208
housefly	12

fruit fly	8
trout	80, 84
chicken	78
rabbit	44
fox	34
horse	64
cattle	60
man	46

You will notice that there seems to be little relationship between the number of chromosomes present and the complexity of the organism. Often very closely related ones will have different numbers of chromosomes. This is one more difficulty to add to our growing list of reasons for not accepting the theory of evolution.

17

What Should I Believe About Prehistoric Men?

Down through the ages men have lived in caves. These range from those of the Indians of western United States to the well-known caves of southern France. Some of the latter still have drawings and paintings on their walls. Many of the animals living at that time are thus pictured. Weapons and tools of various kinds as well as specimens of beautifully carved ivory have been found in some of these caves.

It is unfortunate that when the average person hears the word *caveman* he usually thinks of a stooped, hairy, thick-necked animal with a receding chin and forehead. This is not true of all cavemen; for instance, the Cro-Magnon man. The evolutionist often uses the facts known about cavemen which will emphasize his theory; and so they are usually pictured this way, whether the underlying bones quite warrant it or not. Therefore, because cavemen and evolution are often connected, the creationist avoids the subject as much as possible, thinking that he may be labeled an evolutionist. This is not a scientific attitude, for facts must be faced and explained to the best of our ability. It should not be disturbing if, at times, these explanations have to be changed. As new tests and discoveries are made, new facts are learned.

Until recent years, we had a comparatively short list of the so-called prehistoric men; and some of them were reconstructions from a small number of bones. Some of the best known include the following:

The *Java man* was first found in the Dutch East Indies. The bones are usually said to be about half a million years old.

The *Heidelberg man,* thought to be about half that age, shows many human characteristics.

Neanderthal man is one of the best known, partly because we have so many of his tools and weapons. Many lived in the caves of Europe and Africa. He is usually dated at less than one hundred thousand years, and some scientists say that he lived much more recently than that. He is known to have used fire and likely walked in an upright position. Some of the skeletons have been found surrounded with rings of ceremonial stones or skulls; and some grave sites had meat containers, presumably for use of the deceased in an afterlife. He is one of the early men about whom much is known. This is not a case of having only reconstructions with which to work, for many skeletons have been found. Some of these bones have been dated by modern methods so that we feel quite certain that at least a minimum age for him can be set. However, realize that there is not complete harmony among Christian scientists on this point, but that there are ways of fitting some of these early ages into biblical chronology. They will be pointed out later in this chapter.

The *Cro-Magnon* men were found through many parts of Europe, living probably about twenty-five thousand years ago. It is thought that they took over some of the caves in which the Neanderthal men once lived. These cavemen, in some ways superior to humans living today, had a larger brain size. Their upright posture averaged about six feet in height. Over a hundred such skeletons have been found, so their existence is not a matter of guesswork. Because their physique, maybe even their intelligence, and their impressionistic art compare favorably with that of today, it is difficult for the evolutionist to explain the Cro-Magnon. If amoeba-to-man evolution is true, so also should be the quotations from the high school books which say, "We expect a superman"; "Man is surely continuing to evolve"; and "As we learn more about ourselves we have the priceless opportunity of becoming more human." When Adam and Eve walked and talked with God in the Garden of Eden, we feel that he was the superman and could not have been more human.

When we survey high school texts for accounts of ancient men that have been listed for many years, we usually find mention of the *Peking man.* Many fossils, mostly skulls, were found in China in the early part of this century. The Peking man had many "old" characteristics such as brow ridges and a receding chin and forehead. Although he walked upright and was only about five feet tall, his brain capacity was often as large as that of most modern men. The way in which the long bones and skulls of fossils have been split indicates that he may have been a cannibal.

Other cavemen could be listed, as those found at the foot of Mount Carmel, the *Fontéchevades* finds, and the *Swanscombe man.*

Within the past few decades, many have been unearthed. The best known are the *South Africa man* (Australopithecus) discovered by Broom, Dart, and other scientists. The *East Africa man* (Zinjanthropus) was found in Africa by Dr. and Mrs. Leaky. These fossils and others which have been found in the Olduvai Gorge in East Africa are undoubtedly very old, but we are skeptical about the astonomical dates, ranging between one and two million years, originally assigned to them.

When a fossil is found, how do we know whether it is that of a human rather than one of the animals? In other words, one may well ask, What makes man a man? If we say that some of the cavemen were real men, there must be some criteria for making this assertion. Different scientists would make different lists of human characteristics, but many would include the following:

1. The dental structure is different from that of animals.
2. In the region of the mouth, humans have certain bones to which muscles can be attached allowing the person to speak.
3. Hip and leg bones are constructed to allow the individual to walk upright.
4. To fall within human limits, there are certain body measurements upon which an anthropologist relies. This is especially true of the skull measurements which include brain size.
5. If fossil bones are found closely associated with such things as tools, weapons, and dishes, we usually think of their designer being human. This is especially true if the implements take abstract thought and planning to make.
6. We think of fossils as being human if with them is found any evidence that they worshiped a being which they thought to be superior to themselves.
7. Fossils are considered human if they indicate burial practices showing that they believed in life after death.
8. Man's great toe is not opposable but is in line with the others.

Many other requirements could be listed, but all would not likely be found in any one fossil. However, these are enough to show you that it is often possible to come close to an estimate as to whether the fossil is that of a human being. You can also see how only part of the human skeleton might give this information. We should rule out, though, the overuse of reconstruction—which is a word with which you should be familiar.

When one or more teeth or bones are found, a whole skeleton is sometimes reconstructed from that. If it should be a piece of skull, the rest of the head and whole body might be made of plaster of Paris. In this connection, we will mention the Piltdown man because it has special historical interest. The lower jaw, pieces of skullbone, and a few teeth were found in England in 1912. These fossils always puzzled Christian and nonchristian scientists alike. This was largely because the skull bones seemed to be human and the jawbones that of an ape; yet there were never apes normally living in England. Nevertheless, from these few bones was reconstructed an entire man, and because it was done by an evolutionist, it is natural that he should give to it the form and expression that best fitted his theory. As new methods of dating and study have been devised, they were applied to the Piltdown man in an effort to clear up the mystery surrounding him. The hoax was widely publicized in 1953 when it was established that the jaw had belonged to a different creature from that of the skull. It was shown that the teeth had been altered by an artificial abrasive, stained, and maybe even painted. For what reason these changes had been made and by whom may never be known. It should be noted, though, that the discovery was made by evolutionists, and they were as glad as anyone to have the mystery solved. For this reason, the Piltdown man should not be held to their discredit as is so often done. As we have said before, a scientist must work by setting up a theory, experimenting, breaking it down, and starting again. The story of the Piltdown man should also teach us not to be too dogmatic about single fossils, especially if they are found isolated from others. However, we should not forget that a great deal can be learned about an animal from a single or few bones, as is true with the breastbone of a bird.

From where did the people mentioned in the above paragraphs come? Again there are several theories, but we will mention only one. Accepting completely the inspiration of the whole Bible, we believe that Adam and Eve were the first humans. Only large groups of plants and animals are mentioned in the creation story, and there is no suggestion of details. The story of man is very different. His origin is distinctly set apart from that of animals. Even so, there is little detail so that we do not know what it means that man was made from the dust of the earth. Does it mean that man was created from the same previously created elements which make up dust? This is a possibility. We certainly do not hold to the "mud doll" theory that implies God has a physical body with actual hands, and with them from mud made a human shape, and that He has lungs as we think of them, with which

He breathed actual air. The symbolism may go much deeper than this, and we may not know the details until we reach heaven. In the meantime, we take it by faith and speculate about the details. We see no reason why man should not speculate about the things which are theories. Read the creation account in Psalm 104, and you will get some idea as to how differently the same event may be divinely written.

Closely connected with the age of man is the age of the earth and the universe. It seems safe to say that there are more theories in evangelical Christian circles about these matters of science than there are about evolution in nonchristian ones! Some Christians even hold so tenaciously to their particular theory that they practically say that denial of their theory is a denial of the inspiration of the Word. In forming your theory concerning the age of the earth and of the universe, you must consider also the concept you are going to form about the length of the creative days.

We will mention some of the theories concerning the length of the creative days but not analyze their strengths and weaknesses. One of these has several names, but the easiest to remember is the *gap theory*. People who accept this theory think that there was an original creation, necessarily including a pre-Adamic man. Following this creation, uncertain as to when it began, ended, or how long it lasted, and usually connected with the fall of Satan, they think of a great cataclysm destroying the earth so that it became "without form and void." The account given in Genesis would have been a re-creation when all things were fashioned into the form we see now.

There are people who hold to what is called the *topical order* theory. Instead of the events of Genesis being given in chronological order, they say the order might have been topical. This would be similar to a man building a house and saying, "I built it all: the roof, the walls, and even the fireplace." It is understood that the construction of the house may not have taken place in this order.

Those who believe the *revelatory theory* think that Genesis events may be given in the order in which God revealed them to Moses.

Many people very strongly hold to the theory that everything was created in *twenty-four hour literal, recent days.* This may well be true and you should examine it, but remember that there are other possibilities held by equally evangelical Christians.

Some think of the creation day as being *longer than twenty-four hours.* Even in the creation story, as in many places in the Bible, the word *day* includes a period of time, as in Genesis 2:4 which says, "in the day that the Lord God made the earth and the heavens." Since the Bible does not say how long the creative days were, these people feel that a day could have been a million or even a billion years long. They might or might not have been the same length. There could have been twenty-four hours of creative activity followed by long periods of time, or a long period of time could have had various creative acts scattered throughout it. We do know that Peter says, "one day is with the Lord as a thousand years and a thousand years as one day" (2 Peter 3:8).

We used to think that the age of fossils could be fairly accurately fixed by the rock layer in which they were found. As time has gone by and new methods of dating have been discovered, some of these dates have been confirmed and some have had to be revised. Part of the revision was necessary because there is some uncertainty about the rate at which the rock was deposited and has weathered away or been left at the bottom of the ocean. The whole subject regarding dating fossils by means of the rock layers in which they are found, as well as glaciation, is being reexamined. Maybe *you* will discover something new and helpful. In general, the stratigraphic column still seems to be quite accurate, although there are things about it that we do not understand. For instance, some scientists think that a whole layer of rock may have slid over on top of another, so that the younger rocks are on the bottom and the older ones on the top. This would put the older fossils, many of which bear a number of human characteristics, near the surface. Realize, too, that there are many Christian scientists opposed to this view, but that it might have been possible for such a thing to have happened. In spite of this and other exceptions, in general a fossil found in a rock layer which gives evidence of being 25,000 years old has the same age as that rock layer.

Carbon 14 is likely the best known of the newer dating methods. It is a radioactive substance, and since the carbon 14 in living substances reaches an equilibrium with that in the air, at death all things contain the same amount. As soon as a substance dies, the carbon 14 which it contains starts to disintegrate at a constant rate. Therefore, by using a Geiger counter to measure the amount left in a substance, one can quite accurately tell when the organism died. The more clicks heard per unit of time, the more carbon 14 still remains. One of the limitations of this method is that carbon 14 decomposes so rapidly that it is of little use on any material as it approaches an age of forty thousand years. Another limitation is that carbon must be present in the fossil. Bones lack carbon, so one has to work on the assumption that a fossil is the same age as the wooden tools, weapons, or campfire found near it.

Uranium 238 decomposes so slowly that it is not a good way to date material unless it is older than forty or fifty thousand years. It would be an excellent method of dating if it were not for this fact, and that uranium and fossils are not usually found in the same location.

If you are interested in this matter of dating ancient fossils, you should get some science books and study the potassium-argon method and others such as those involving fluorine or gypsum, as well as the one called fission-tracking. If at all possible, when dating a fossil we try not to depend on any one method. However, if two or more methods give about the same date, we feel that it is more meaningful.

We have mentioned various so-called prehistoric men and a number of dating methods. The length and purpose of this book do not allow us to analyze them. However, we feel that there is one comment we should make before finishing this chapter. We think Adam was created long before the 4004 B.C. date figured by Ussher in the 1600s. Most scientists think that at least some of the ancient men, such as the Cro-Magnon and Swanscombe, were much older than Ussher's date and yet were real human beings. The date 4004 B.C. is still accepted by many people because it was done near the time the King James Version of the Bible was translated, and by some is considered as inspired as the Bible!

If one were to accept Ussher's date, it would necessitate accepting the assumptions upon which he did his chronological work. Since the seventeenth century, study of the Bible and the facts of science, as those in archeology, have led us to believe that some of his work was not done on a sound basis. For Abraham and following events, Ussher's date of 2,000 B.C. still seems accurate. There are a number of matters concerning the time from creation to Abraham about which people disagree.

Genesis 5 is used to arrive at dates from Adam to Noah, and Genesis 11 for dates from Noah to Abraham. However, we know that in ancient Israel some of the genealogical words which we use were then used in an entirely different way. For instance, in Genesis 46:16-18 the word *son* has four different meanings. It means an actual son as we think of it today, a grandson, a granddaughter, or even a great-grandson. Matthew 1:1 illustrates how the use of the words *son* and *generation* have changed from Bible times to the present, for it says, "The Book of the generation of Jesus Christ, the son of David, the son of Abraham." This selective style of writing is consistent with God's purpose of writing to a specific people at a particular time. In the creation account it would be purposeless to name individually each of the hundreds of thousands of individual plants and animals that He created. When the Genesis account mentions herbs, grass, and fruit trees, we take it for granted that He also created the mosses, ferns, and other plants. This same kind of selection seems to have been used in other genealogies. Four wicked kings are left out of the genealogy of Christ as given in Matthew.

Throughout the centuries, the use of the word *begat* has also changed. At that time, it did not always mean to give birth to a son or daughter. Instead, it meant to become the ancestor of a certain person. It could mean not only a son or daughter, but was sometimes used for the head of a whole nation or group of people. Genesis 10:15-18 lists eleven people begotten by Canaan. Two of these were sons, and nine were families or nations. The account of the life of Abraham is a good example of the change in the use of the word begat. Genesis 11:26 says that Terah lived seventy years and begat Abram, Nahor and Haran. There is no indication that these three were triplets born when their father was seventy years old. Look up and compare this verse with Genesis 11:32 and 12:4. It would appear that Abram was born when his father was one hundred and thirty, but that when Terah was seventy years old, he had been designated to be the ancestor of a man who would be born sixty years later.

There is another reason why we should not assume that we can figure dates from the creation to Abraham by adding the number of years given for the life of the men listed. Some of the genealogies have certain names or a number of generations omitted. In this connection, compare the genealogies given in 1 Chronicles 6 with those in Ezra 7.

Changes from our present usage or omissions in genealogies should in no way alter our realization that the Bible is the Word of God. Selection had to be used and we are told that if all that the Lord did and said while on earth were written, the world could not contain the books. The purpose of the first part of Genesis is not to give us information about every man mentioned, but to let us know that God created all things. He quickly moves on (as soon as the third chapter) to give a promise of the coming Redeemer who would save those who put their trust in Him from the sin which had entered the world bringing death with it. In view of the real purpose of the Bible, it seems of little value compared with other things to spend much time on chronologies.

Never lose sight of the fact, however, that the genealogies are as much inspired as the rest of the Word. The inspired Word and the created world both originated in a God of truth, and it is only when there is a misinterpretation of one of them that there may seem to be a lack of harmony. We know that this harmony exists, but we also know that it sometimes takes real study to discover it. Therefore, be willing to study.

18

How Will My Decision About Evolution Be Directive in My Life?

YOU ARE A RESPONSIBLE HUMAN BEING. You are not a creature who is the result of an evolutionary process. You belong to a race of people created in the image of God. Your ancestors were not apes. Your origin was not the result of a chance combination of chemicals. Your ancestors never had to crawl in the dust or swing through the trees of a forest. They were men and women who had God-breathed life within them. Adam and Eve, your first ancestors, were intelligent, upright beings. As they came fresh from the hand of God, they were untouched by sin and its results. They were a creation separate from animals, although made from the dust of the ground whose elements God had already created.

God made man perfect with a body, soul, and spirit. He had the full ability to understand spiritual ideas and to love and glorify God. The spiritual image of God given to man was marred by sin. Ever hear that word *sin?* It's almost extinct in some vocabularies and concepts. Well, it's still here and as real as ever, "For all have sinned, and come short of the glory of God" (Romans 3:23). Ever since man decided with his God-given will to disobey God, there has been sin with its inescapable results, "For the wages of sin is death" (Romans 6:23). You can't explain away or shut your eyes to death which came into the world with sin. Both physical and spiritual death came to trouble the human race. Death is still here. It may come slowly with old age and disease. It may come suddenly and unexpectedly with tragedy and horror. Instead of death, you want to live fully and freely. You want to have fun, but your real desire is for satisfaction now and forever. Mere fun doesn't satisfy.

Spiritual death is already your companion if you haven't recognized the God who created the human race as the one person in the universe to whom you must yield yourself. The reason you are responsible to God is that He made you able to think and reason, love and hate, ignore or acknowledge Him. His purpose for creating you was that you would love and glorify Him with all of your being. You have this ability and capacity. He did not give this spiritual capacity to animals. The cutest and most intelligent little monkey can be taught to kneel on his hind legs, fold his front paws, bow his head, and close his eyes; but he cannot pray. You can. All humans possess a God-given consciousness of God. They realize that a supreme being exists.

Nature declares it to them. "For the invisible things of him from the creation of the world are clearly seen, being understood by the things that are made, even his eternal power and Godhead; so that they are without excuse" (Romans 1:20). The universe tells by its order, majesty, beauty, and design that a Person called it into being.

The mind questions the origin of the human being, the earth, and the universe. The small child says, "Mommie, where did I come from?" The young adolescent often wonders, "Am I the real child of this man and woman who claim to be my parents?" The teenager has been expected to believe in evolution ever since he entered public school. Even though reared in a Christian home, he may well ask honestly, "Who is right, my church and parents—or my textbooks and teachers?"

Why do these questions continue to bother the human mind and demand an answer? They are part of belonging to a created group different from any other. They are the result of possessing not only a body or even an ability to do some thinking, loving, and deciding, but also of having the capacity to know that a supreme personality exists and operates in the universe. As a result, man turns to worshiping something. It may be stones, trees, the sun, spirits, ancestors, idols, strange creatures of his own imaginings, or the doing of good works. Blood sacrifices may be offered to appease evil deities—even children are offered. Some inflict self-torture. None of this satisfies. The intellectual who wishes to avoid this soul-condemning search for a solution to his heart hunger turns to science to find a possible answer. He does not want to consider himself a sinner in the eyes of a righteous God, for this would include a belief in the need of retribution and would rob him of his independence. The theory of evolution, of believing that all nature is improving, that all animals are adjusting to environments and growing more complex and intelligent, and that man is on an upward course toward perfection seems to solve the problem—temporarily and partially. But man's intellect never warms his heart nor leads him to God.

Think of yourself as a triune being composed of body, soul, and spirit. Your body contacts the world about you through the eyes, ears, and other senses. Animals have similar and wonderful systems of circulation, respiration, and reproduction. Your soul cannot be seen and has no substance, yet it is very real. With it you think, express your emotions, and exercise your will or power of choice. An animal has some of these abilities in a limited and temporary fashion. When its body ceases to exist, so does its soul. Not so with man.

The spirit of man can know and love God. It is because man has a spirit that he can desire to know God. The spirit, darkened and separated from God when man first sinned, is "quickened" or made alive when a human soul has faith in Jesus Christ. "Believe on the Lord Jesus Christ, and thou shalt be saved" (Acts 16:31). He is the *only* way, for

He said, "I am the way, the truth and the life" (John 14:6).

Now is the time. The *now* generation, even though exterior manners and appearances change, has the same heart cry. It can be satisfied only in the same way. Look to Jesus now. The shackles of customs, rituals, formalities, traditions of family, church, and society will fall away as you see and confess the Christ of God. He will make you a *new creation.* This is the creation that really matters to you. Even as God spoke the worlds into being, He can now speak new life into you, if you will listen to His voice. When you accept Christ's work in your behalf, an eternity of knowing God begins.

Will you accept Him *now?*

Appendix

Biology Texts

SEE THE ACCOMPANYING CHART showing where the teaching concerning the theory of evolution is found in some high school texts. The books analyzed are among the ones most commonly used in the United States. If you study from one of these books, or have access to one to read, you will be interested to look at the suggested pages. They will give you the arguments in various fields, proposed by the evolutionist to account for his theory.

Ten years ago it was a comparatively easy task to compile such a chart, for most of the information about the theory was contained in one chapter. This was usually located near the end of the book. Today, even the revised books by the same authors, have undergone a radical change. In most of the texts, the largest amount of material is still contained in one chapter. However, scattered throughout the entire book and woven into its very basic framework, one also finds much more of the same teaching. Here and there throughout the books are implications, sentences and paragraphs about this theory. This is generally to the effect that "all reputable scientists" have accepted it as a theory that most facts support and against which there is very little evidence. This

This section is from the author's earlier book, ***Evolution on Trial*** (Chicago: Moody, 1970), pp. 182-84.

WHAT VARIOUS HIGH SCHOOL BIOLOGY TEXTBOOKS TEACH

Textbooks	*Similarity as Evidence for Relationship*	*Evidence from Geology and Paleontology*	*Change in the Horse*	*Evidence from Classification*
	Pages	Pages	Pages	Pages
Biological Science by W. H. Gregory and E. H. Goldman	707-8	705-6; 710-17	448-49	91-102; 706
Biology by W. J. Kimball	12-16; 543-45	540-43; 581-83		12-18
Biology by C. A. Villee	212; 610-11	592-609	607	92-93; 211-13; 610
Biology by E. Kroeber, W. H. Wolff, and R. L. Weaver	480-86	460-80; 485	476; 478-99	127-37; 483-84; 486
Biology for You by B. B. Vance and D. F. Miller	524; 534-35	525-35	530	210-11; 531
Design for Life by R. E. Trump and D. L. Fagle	433-38	425-27; 441-44	414-15	431-37; 446-52
Elements of Biology by W. M. Smallwood, I. L. Reveley and G. A. Bailey. Revised by R. A. Dodge	627-34; 661	619-25	623-24	603-16; 627-30
Modern Biology by J. H. Otto and A. Towle	182; 184; 542	182-83; 543-49	522	199-207

CONCERNING THE THEORY OF EVOLUTION

Biochemistry	*Geographic Distribution*	*Embryology*	*Vestigial Structures*	*Origin of the Universe*	*Origin of Life*	*Man*	*Genetic Cause of Change*
Pages	Pages	Pages	Pages	Pages	Pages	Pages	Pages
709-10	450-52; 706-7	708-9	708	9-10	10; 710	464-66; 471-85	721-33
546-48	550-53; 565-91	544-46	545		576-81	612-15	539-40; 548-50; 555-73
611-12	605-6; 615-16	612-14	611		587-89	618-33	575-91; 614-15
453-86	484-86	482-86	483; 486	460	362-64	495-505	488-93
	523-25			525	529; 533	535-40	489-518; 539
442-45	419-25; 445-46	438-40	439-41		45-46	607-21	406-30
	630	625-27	631	618	618-19; 622	266-77	582-99
204; 542	190-97	182-84	182; 184		18-25	542-49	184-90; 193

book has shown you that the theory has many weaknesses. Existing evidence may be used to support the idea of divine creation as it is revealed to us in the Bible.

Not included in the chart of texts are four high school biology books about which we would say an extra word. These are: *The Science of Biology* by P. B. Weisz; and *Biological Science: An Inquiry Into Life*, Yellow Version; *Biological Science, Molecules to Man*, Blue Version; *High School Biology: BSCS* Green Version.

The last three books were produced by the Biological Science Curriculum Study committee and are usually referred to as the BSCS books. Experts in many fields worked together for a number of years to produce these attractive texts. Many people feel that they are superior to those that a single author could produce and they have been widely adopted by school systems.

The three BSCS books, as well as the one by P. B. Weisz, are so completely evolutionary that it is impossible to analyze them as has been done with the other texts. Not only is organic evolution taken for granted, but it is the very framework of each book. It permeates each text so that no attempt has been made to give the pages on which evidence for the theory is presented.